CRYSTAL ENERGY

CRYSTAL ENERGY

Put the Power in the Palm of Your Hand

GARI GOLD

CONTEMPORARY
BOOKS, INC.
CHICAGO • NEW YORK

Library of Congress Cataloging-in-Publication Data

Gold, Gari.
Crystal energy : put the power in the palm of your hand / Gari Gold.
p. cm.
ISBN 0-8092-4640-6 (pbk.)
1. Crystal-gazing. I. Title.
BF1331.G64 1987
131—dc19 87-17782
CIP

Photography by David Sorcher

Published by Contemporary Books, Inc.
180 North Michigan Avenue, Chicago, Illinois 60601
Manufactured in the United States of America
Library of Congress Catalog Card Number: 87-17782
International Standard Book Number: 0-8092-4640-6

Published simultaneously in Canada by Beaverbooks, Ltd.
195 Allstate Parkway, Valleywood Business Park
Markham, Ontario L3R 4T8 Canada

Dedicated to you,
the one who made this book possible.
Thanks.

The material herein is intended as information only, and is presented without any claim of scientific proof. The author's intention is not to prescribe cures for various conditions, but merely to present alternative healing methods from our own earth that can be used with other healing therapies.

♦ CONTENTS

♦ ACKNOWLEDGMENTS

I'd like to acknowledge the many sources of information and inspiration that went into the creation of this slim volume. The many people I have met through crystal work have provided input and confirmation of intuitive knowledge, for which I am grateful. I'd like to thank the many teachers who have freely passed their knowledge to me, as well as the healers who have lovingly touched my life and helped me make many positive changes. And I'd like to thank my close friends, who have encouraged and supported me in following my heart. I especially thank Jen, who always has an ear, and usually a joke; Tony Locane, for his help in editing the material for this book and for his advice, which comes from such an open heart; and all the people who have trusted me to guide them, for this is where I receive my greatest learning experience. And last, I thank the crystals, meditation, music, and the vision of a beautiful future, as my greatest hope and inspiration.

CRYSTAL ENERGY

♦ INTRODUCTION

When you bought this book or received it as a gift, you received with it a quartz crystal.

You may have heard that a crystal is a sort of trendy, New Age rabbit's foot and that everyone is carrying one in his or her pocket these days. While crystals *are* enjoying a lot of attention, they are not merely good-luck charms or pretty trinkets; they are powerful tools to tap into your internal energies and into the natural world around you.

Great numbers of crystals exist beneath the earth's crust. Many who work with crystals feel that crystals help keep the earth's magnetic field flowing and in balance. Crystals, although mildly magnetic, seem to flow with and have flowing through them a quality that is similar to electricity and to magnetism but that is neither.

Like electricity and magnetism, crystal energy is a natural force that we do not fully understand. We know many of the characteristics of other kinds of power, like electricity and magnetism, and accept these forces in our lives. In the same way, we can accept the potential of crystal energy and learn more about it.

In 200 years, we will look

back at books like this one and feel that we are reading a "Dick and Jane" first-year primer. And indeed we are. There is really no knowing where our experimentation with crystals will lead. Our present efforts are a bit like opening the tiny cap on the giant economy size tube of toothpaste. But as you gradually learn the story of quartz, the use of quartz in human history, its uses in technology today, and the rising interest in quartz among healers and lay persons alike, you will probably find some answers about why you were attracted to this book and, specifically, how this particular quartz found you, or vice versa. Really, we are all writing this book together.

Gems have been used off and on throughout history, but never before so widely or with such curiosity, and acceptance. Probably the use of crystals will continue to increase as we approach the Age of Aquarius. Awareness of the power of stones is growing so that we can help ourselves to become healthy and strong, and have the strength and insight to run our lives well for ourselves as individuals, and collectively as humanity.

The quartz crystal, although not a conductor of electricity, seems to function as a conductor of subtle, or psychic, energy.

1 ♦ GETTING TO KNOW YOUR QUARTZ CRYSTAL

Place your crystal near you, so that you can look at it throughout this discussion of the past, present, and future of quartz crystals very much like the one you now possess. Look at your quartz crystal—hold it to the light, look through it, and give it a good general inspection if you haven't already.

Quartz crystal is also referred to as clear quartz or sometimes simply crystal.

Don't confuse quartz "crystal" with glass or "lead crystal." Glass is manufactured and often contains lead in small percentages to give it weight and help keep it colorless. However, glass crystal is made mostly of what geologists and scientists call silica, another name for quartz. Therefore, glass crystal and quartz crystal look similar.

So, what's the big difference?

WHAT ARE QUARTZ CRYSTALS?

If there is anything we have more of on planet earth than silicon, it's oxygen. When the two come together, they form a new molecular structure, SiO_2, chemistry-ese for silicon dioxide, also called silica. Silica is the most abundant mineral on our planet—if the earth were a cake, silica would probably be the flour. In its noncrystalline form, in fact, silica is the basis for sand, which we melt to make glass. In its crystalline form, however, silica becomes quartz.

Crystals of silica were created by the slow cooling of molten silica as the earth formed and cooled and the hotter inner liquids shifted and bubbled up toward the cooler outer crust. As the molten silica was pushed to the earth's surface, it changed structure and form. Voilà—crystals were born.

Your stone actually developed naturally, under the surface of the earth, long before we were born—for that matter, long before our parents and long before the birth of our country or civilization as we know it. Our civilization is only a few thousand years old. Your quartz has probably been around for hundreds of thousands of years. And it was formed in exactly the shape in which you see it; *it has not been cut or faceted*. No person or machine has altered it to create the shape you now hold in your hand.

You have an example of quartz crystal structure. They all look pretty much alike. Every molecule (the smallest form in which a substance can exist) of quartz looks just like the whole crystal. Or, otherwise stated, the crystal is an external reflection of the internal arrangement of its molecules. This means that the arrangement of the atoms that make up your crystal looks exactly like that very same crystal. Your crystal is a macrocosm of its molecules.

Look at your crystal again, and count the long flat sides. There should be six. If you counted more or less, count again, this time more carefully. It's easy to make a mistake because the planes look pretty much the same. You'll get better at seeing the differences, the more you look.

Notice how these long planes angle off at one end and then come to a point, also called a "termination." Just as your crystal has six sides, it has six triangular planes that make up the termination. Take a minute to count the planes that make up the termination and to become familiar with the way in which each long plane bends into each of the corresponding triangles of the six-sided termination. Three of the triangular sides are probably smaller than the other three, and the place where all these planes come together is not necessarily equidistant from all the sides. It's a lopsided six-sided pyramid, and the tip of the pyramid may not be exactly a "point" like a dot; it may actually come together like a line or a razor edge. Don't worry if your point is chipped; one of the amazing things about crystals is that

Crystal cluster.

they function well without a perfect termination; you can reconstruct their original shape in your mind's eye.

Now look at the end opposite the termination. In the case of your crystal, this end is broken. It was formerly part of a larger group of crystals called a "cluster." In a cluster, the crystals are massed up against each other, often infringing on the space of one another, slightly altering each other's growth. You may sometimes notice a notch on a crystal that indicates where another crystal was when they grew in a cluster. A group of crystals in a cluster tend to resemble each other, as family members do.

The broken end is whiter and cloudier, while the tip is much more transparent. There are a number of reasons for this, but for now, let's chalk it up to chemistry and the nature of things and how they grow.

Some quartz crystals are "double-terminated." As you might well imagine, these crystals look like a barrel with a six-sided pyramid on each end. Although they are not uncommon, this book will describe single-terminated crystals unless otherwise specified.

One final word about words. To be less formal, I'm going to drop the term *quartz crystal* in favor of the simpler use of either *quartz* or *crystal* alone. These terms are intended to refer to objects like the one you are holding. But you should recognize that these terms, while accepted, are not entirely correct. First, not all quartz forms into crystals. Furthermore, not all crystals are quartz. Almost all gemstones, and many minerals not considered to be pretty enough to be worn as jewelry (as well as metals, under the right conditions) form into crystals.

WHERE DO QUARTZ CRYSTALS COME FROM?

Quartz is the most abundant mineral on earth, and it stands to reason that quartz crystals can be found all over the earth. They're found high in the peaks of the Swiss and

Italian mountains and throughout the French Alps. The marble quarries in Italy also contain plenty of quartz, this being generally of excellent quality. The rich deposits of Madagascar were mined extensively through the eighteenth century and are still producing great quantities of high-quality stone. Rock crystal (still another name for the same material) is also found in India, the United States, Canada, and Mexico. Brazil has large amounts of clear quartz, as well as the colored varieties, which we will cover later.

Quartz mines are scattered all over the United States. A beautiful quality quartz, for example, can be found in

Double-terminated quartz crystal.

Herkimer, New York. These "Herkimer diamonds" are called diamonds because of their amazing transparency. Also, quartz mines in North Carolina, California, and Arkansas are currently enjoying a lot of good press. If you are particularly interested in visiting a mine, many are accessible to the amateur on vacation, and you can combine camping with digging around in the hills and come away with crystals that you have found yourself. Many people, myself included, are content to limit our digging to the many wonderful rock and mineral shops that have sprung up all around urban areas.

Try to see in your mind's eye an image of the earth. See it turning slowly, blue waters, green and brown patches, clouds dotting the surface. Bring yourself a little closer to the surface, so that you see the streams trickle off into large rivers and then oceans. See the mountain ranges. See how they help define where the water flows and where it doesn't flow. Follow the patterns of the mountain ranges in your mind. This is beginning to look like a lacework pattern, mountains and water, streams intertwining. Now put on your Superman Laser Vision and see under the surface of the earth. See quartz crystal running in what geologists call veins.

These quartz veins add another dimension to the lacelike complexities of elements connecting and drifting apart and reconnecting on the crust of the earth. Occasionally, these veins create small "lakes," referred to as caves, and then the veins continue on, perhaps diminishing, disappearing and reappearing. Running closely with the quartz veins are usually veins of other gemstones, and perhaps a pocket of gold or platinum. Veins of rarer stones, such as emeralds, are scattered about under the surface of the earth, and are more difficult to locate than quartz.

Much rarer than emerald is tanzanite, a beautiful pale blue stone, which is found only in East Africa. Pink tanzanite (zoisite), is found in small quantities in Norway and in North Carolina—this gem is barely sprinkled throughout the world. Can you find another pocket of tanzanite in your mind's eye? Could it be somewhere in Asia or somewhere near the mine in North Carolina but just a little deeper into the earth than anyone bothered to dig?

Now take your imagination back to the blue and green/brown surface of the earth and refocus your attention on your own crystal. It is not a gift from me, or the publisher, or your Aunt Susie: It is truly a gift from the earth. Your crystal is unique. That means there is not another one exactly like it on the planet.

Quartz is found almost everywhere you would think of looking and in many places you wouldn't. The family of quartz includes many interesting members. The next section introduces you to the variety.

Crystals can:

- **Enhance the quality of your meditation or prayer.**
- **Promote a feeling of having more energy.**
- **Assist efforts to increase your psychic abilities and extra sensory perception (ESP).**
- **Enhance your health.**
- **Enhance the dream state and help you remember dreams.**
- **Relieve localized pain.**
- **Help you create a desirable change in your life.**
- **Increase your ability to communicate by helping you develop your abilities as a sender and receiver of thought vibrations.**
- **Open your universal vibrations.**

VARIATIONS ON A THEME OF QUARTZ

Amethyst

It may surprise you that amethyst is of the quartz group. It is considered a semiprecious stone by the jewelry trade and can often be found with exceptional color quality. It ranges from a deep and brilliant purple to the palest of lilac. The quality of amethyst (and most gems) is judged to be good when the color is fine and evenly distributed within the gem. Other qualities, such as rarity, are figured in to determine jewel price. Amethyst was once considered the stone of royalty, and was equal to diamond in desirability. The amethyst has long been considered an amulet to protect against drunkenness.

Crystals cannot:

- **Take out the garbage.**
- **Return your phone calls.**
- **Fix it with your spouse when you forget important dates.**
- **Relieve you from taking responsibility for your own life. Crystals are *not* a substitute for any type of medical, holistic medical, or psychological attention that you feel might be needed. The type of work that is done with crystals as tools is *not indicated in every situation* nor is it necessarily recommended in every case as the best avenue of therapy. *Do not hesitate* to seek advice from a trusted professional therapist when you have a health question.**

Citrine

Also considered a semiprecious stone and often cut for jewelry, citrine's name suggests its lemon-orange range of color. Citrine can look so similar to more expensive gemstones like topaz or yellow beryl that it is often sold to an uneducated buyer under some fancy name like "Brazilian topaz" or "smoky topaz." For a cut stone, often only an expert can tell the difference. As an uncut or rough stone, citrine is easy to recognize. It is similar to clear quartz in that it has a six-sided pyramidal termination. The plane sides, however, are very short, seemingly nonexistent in some specimens.

Amethyst, citrine, and clear quartz often occur as geodes, round, usually hollow balls of rock that are lined on the inside with crystals. These are common but lovely when sawed in halves and polished.

Smoky Quartz

Another quartz that gets its name from its appearance is smoky quartz. Often, smoky quartz's appearance is yellowish or brown or even gray. The depth of the color ranges from a light tint to a density that is almost opaque. The shape of smokies resembles that of clear quartz.

Inclusions in Quartz

Inclusions in a crystal are exactly what they sound like. They are minerals or metals that are "included" in another mineral. Clear quartz and smoky quartz commonly have inclusions. These can be very beautiful and can enhance the healing aspects of the stone.

The most common inclusions are tourmaline and rutile. *Tourmaline* inclusions are typically black or green. As an inclusion in quartz, *rutile* (TiO_2 or titanium dioxide) looks golden and darts across the quartz in straight, fine, hairlike crystals.

Rose Quartz

Most often, rose quartz is found in an amorphous or massive state. This means that as the silica cooled, it did not form large individual crystals, but developed microscopically small, or microcrystalline, crystals.

Chalcedony

The term *chalcedony* refers to any of a large group of stones that, although they are chemically the same as quartz crystal (SiO_2), look rather

Rutile sphere.

different, usually colored and in the microcrystalline state. When cut and polished, chalcedony is a highly desirable gemstone for jewelry. Many stones in this group are called "agate."

Types of Chalcedony

- **Aventurine** *is a green stone that sparkles because it has mica as inclusions. It is also found in yellow and blue.*
- **Blue lace agate** *is a powder blue and is sometimes veined with white.*
- **Blue or green quartz** *resembles rose quartz, except that it is blue or green. It is fairly rare. Interestingly enough, the green variety is found near deposits of emerald.*

- **Carnelian and sard** *are earthy red stones. Sard is brownish and the more orangy; translucent colors are carnelian.*
- **Cat's-eye, or Occidental cat's eye**, *is green or yellowish brown, and has a common optical effect called chatoyancy, which is due to an asbestos inclusion. When cut and polished, this stone shows a reflection of light in a milky-white wave across the stone. This is thought to resemble a cat's eye.*
- **Chrysoprase** *is apple green and is often used for jewelry because of its fine color.*
- **Flint** *is whitish, or a dull gray or black, but it can be brownish red. You won't be wearing this stone, but chances are good that it's setting sparks in your lighter.*
- **Heliotrope** *is also known as* **bloodstone**. *It is dark green and spotted with red inclusions.*
- **Jasper**, *another earthy red stone, is often streaked with black or white.*
- **Moss agate**, *also called* **dendric quartz**, *looks like it has moss captured inside. It doesn't. This mosslike inclusion is actually another mineral, called hornblend.*
- **Onyx and sardonyx** *are streaked agates. When they are cut thinly along the stripes of colors and carved, you have what is called a cameo (seen often in carnelian, too).*
- **Picture agate** *looks like a landscape when cut across its layers of colors and polished.*
- **Tiger's-eye** *is yellow- and black-banded quartz. It seems to shimmer, like a tiger's eye.*

Silicon dioxide (quartz) is found on earth in many forms besides as gemstones called quartz and chalcedony. Our beaches are quartz, and many of the common rocks we see around us are formed of many different minerals including quartz. For example, obsidian, glassy and usually black, is volcanic rock and is more than one-third quartz. Granite, a common rock, contains varying amounts of quartz.

2 ♦ THE ROCK OF AGES

Looking at the past is always a little like putting the pieces of a puzzle together, with a few pieces missing. Eyewitness accounts are difficult to come by. Over the past few hundred years, archeologists have dug their fingers to the bone discovering all sorts of things in order to enrich our understanding of our past. Every civilization has had its own uses for gemstones. We may find many of the old uses of stones "superstitious" today, but without certain of the technological advances we enjoy, we would be more understanding of the importance of adornment and ritual in the lives of our ancestors. I know that if I didn't have nonstop running water in my apartment I'd be out there shaking my rattle and dressing up with the best of them, and offering anything to the sky I thought might help tempt the production of rainclouds. It couldn't hurt, and it might even keep my mind off being thirsty.

STONES BEFORE HISTORY

Let's travel back in time for a moment or two, leaving behind civilization as we understand it, and think about our paleolithic roots. Daily

The "crystal" covering your watch was once made of just that. But the first uses of quartz high technology were lenses. The first lenses were ground from quartz crystals and used in the early telescopes. Quartz is still used for lenses in some telescopes, microscopes, and some other optical devices.

life about 20,000 years ago was quite different from now. Survival depended on gathering whatever food was provided by nature and by hunting small game. This period is called the Stone Age because of the stone tools formed then, which are our primary source of information about these people. These stone tools were the first man-made tools. During this period, people were completely dependent on nature, and the first religions were based on ensuring that there would be enough food to gather.

Early on, every member of the community was involved in ritual ceremonies, but toward the end of the paleolithic period, shamans or medicine men appeared on the scene—only part-time, though, because in a food-gathering society, every hand is needed. Today, the food-gathering cultures of Australian aborigines, Bushmen, Eskimos, and American Indians all have shamans or magical specialists. In each culture, the shaman uses crystals and stones in his or her medicine bag. We assume most "primitive" peoples did the same.

Slowly, societies made the transition to domesticating animals and farming, a way of life that was more efficient because a few could do the work to feed many. The shaman now could be left to the magical arts, and there was plenty to do, for now the culture needed spirits to

Yes, Virginia, there really is a quartz in your wristwatch. The quick rundown on that facet (!) of the quartz story is as follows: Quartz is used in a watch to regulate the energy from the battery that runs the gears that run the hands on the watch. For every certain number of oscillations of the quartz, the gears click that a second has passed. The piece of quartz in a watch is measured for its frequency; if it oscillates too slowly, it can be cut smaller so that it will vibrate at the desired rate. In fact, stones have always been an integral part of precision timepieces.

watch over flocks and fields. It was at this time that the fertility goddess was born.

People wore or carried amulets and talismans. Amulets are of a protective nature, and talismans are a remembrance of some sort of spell or initiation. Stones found on the surface of the earth or found in shallow streams were probably bound together with animal skin, feathers, or plant materials.

These amulets and talismans may seem pretty silly and superstitious to us now, but I assure you we continue these traditions even today. Wedding bands and engagement rings are common talismans. There has been an initiation into engagement and marriage. The diamond engagement ring is also a symbol of purity and duration. Class rings, often displaying stones, are symbols of initiation or of belonging to a certain group; so are fraternity and sorority pins.

Amulets are less important in our contemporary culture because the threat of charging tigers seems pretty removed from our consciousness, but how many businessmen have “lucky ties” or certain sweaters that are sure to clinch the deal at the meeting? Mezuzahs are posted on many doorposts, but we rely more on our electronic amulets—the burglar alarm and the smoke detector.

STONES THROUGHOUT HISTORY

Stones have been used throughout history as medicine and in relation to matters of the spirit. The ancient Egyptians, who have furnished us with greatly detailed information of their

Quartz exhibits a characteristic called piezoelectricity. This complex phenomenon, involving almost every aspect of classical physics, was discovered by the Curie brothers in France only about one hundred years ago. Briefly, if a slice of quartz is mechanically compressed, it becomes electrically charged, positive on one side and negative on the other, which allows an electrical current to be passed through the crystal. This phenomenon has inspired great changes in modern technology. A slice of quartz oscillates at a precise rate, which is unaffected by a wide range of temperature changes. This is important in building communication systems, time-keeping devices, and many computers.

Quartz crystal is used in some computers to regulate oscillation elements in much the same way it works in a quartz watch. But don't confuse quartz (silicon dioxide, SiO_2) with the silicon (Si) that is used in the manufacture of semiconductors. Semiconductors are the amazing little printed circuits that have reduced the size of computers from roomsize to small enough to fit on one's desk.

daily life and their religious rituals, provide the first concrete evidence of the use of gemstones for health and magical powers. Amulets, and the stones they were fashioned from, were very important to the Egyptians, to the living and to the dead. Lapis lazuli, carnelian, feldspar, and serpentine (jadelike) were the preferred stones. Gold was the preferred metal, but copper, bronze, and iron were also used frequently. Most amulets were made of faience, a fine claylike paste made from ground quartz or sand. They were baked—for hardness, and to bring out the glaze—in molds the shape of the eye of horus (ensuring good health and regeneration), the ankh (the symbol of life), and, most popular of all, the scarab, which the Egyptians felt assured existence. These quartz-paste amulets were all made with holes so that they were easily strung and worn.

Stones were often ground up, and used as medicine and ointments or actually ingested. There are written accounts of these medicines, but no accounts of what happened to the unfortunate patients. Many stones are poisonous if taken internally—don't even think of experimenting with swallowing stones, ground up or whole! Malachite and azurite were also ground and used as cosmetics, hardly a hypoallergenic practice.

It is said that Cleopatra wore a headdress and a belt of hematite, which was supposed to help her retain her youthfulness and her historically infamous fabulous face. Headdresses (crowns) were also found that had been lined with malachite, a stone that the Egyptians used to restore poor eyesight. Perhaps these crowns were lined to give the wearers "vision" in making wise decisions.

Further evidence of the spiritual use of stones can be found in the Bible. For example, Exodus 28:15-21 reads:

> Fashion a breastplate for making decisions—the work of a skilled craftsman. Make it like the ephod: of gold, and

Quartz crystals are often cultured—grown in the laboratory—because natural quartz often has imperfections that make it unsuitable for specific technical purposes. Cultured quartz can be used for most applications, although natural quartz may be preferred in some cases.

of blue, purple, and scarlet yarn, and of finely twisted linen. . . . Then mount four rows of precious stones on it. In the first row there shall be a ruby, a topaz and a beryl; in the second row a turquoise, a sapphire and an emerald; in the third row a jacinth, an agate and an amethyst; in the fourth row a chrysolite, an onyx and a jasper. Mount them in gold filigree settings. There are to be twelve stones, one for each of the names of the sons of Israel, each engraved like a seal with the names of the one of the twelve tribes.

Beginning at verse 28, the high priest Aaron is instructed to wear the breastplate upon entering the Holy Place before the Lord and to "put the Urim and the Thummim" (two unidentified objects) on the breastplate "so that they may be over Aaron's heart whenever he enters the presence of the Lord. Thus Aaron will always bear the means of making decisions for the Israelites over his heart before the Lord." (28:30.)

This passage of the Bible is important in the discussion of historical uses of gemstones, and is often referred to. The breastplate of Aaron seems to be particularly important, as do many of the details described in the Bible for creating the correct atmosphere and ritual objects for the place of communion with the Lord.

It is interesting to note that the stones in Aaron's breastplate cover his heart, which, as we'll see later, is one of the body's main energy centers. That the stones should cover Aaron's chest during the decision-making process and communion with the Lord suggests that communication with divinity takes place through the heart, or that one's wisest decisions are the ones arrived at through the heart.

Also important is the parallel between Aaron's and subsequent high priests' breastplates and the traditional king's crown as a symbol of power. Both wearers were deemed to have some sort of divine instruction. Clearly, the stones were believed to play an important role in this communication.

The Europeans of the medieval period made a great showing of gemstones for amulets and talismans. They also ground up stones as medicines and attributed specific functions to the many stones that were mined and traded along the expanding trade routes.

Crystal balls also come to mind; these spheres of rock crystal were used primarily in Europe over the last few centuries by psychics and

In its natural or unprogrammed state, the frequencies on which quartz transmits give the user a sense of increased awareness and ability to focus one's attention. This sense tends to increase when the crystal is held in the receiving hand (left hand for righties) or otherwise in contact with the user's skin. Quartz can also be programmed; it will then emanate back to you what you have programmed with your intent.

clairvoyants as a method of looking into the future, or of looking into another space at the present time, or of gathering information from the past. This method is called scrying. The original technique of scrying involved specially prepared mirrors or a glass bowl filled with water—the technique could even be done with a pool of ink. It was felt that the quartz was of a material that helped the clairvoyant access the material to be "seen." Some modern-day psychics still use crystal balls.

There are many reasons why stones have fallen into the category of healing aid or talisman, or been viewed as merely pretty baubles, or been used as a measure of wealth. (Why *are* stones considered valuable, anyway? It certainly makes *me* think.) These questions are complicated and related to science, social science, and the growth of religious leadership. And the answers really have little bearing on the present moment, or on you, your quartz crystal, and what people are currently discovering about gemstones in the present or what they will discover in the future. We are on the edge of a new age. I just thought some history would let you know that the new age is tied to the wisdom of the ages.

3 ♦ ENERGY AND YOUR LIFE FORCE

Before you start using crystals to direct energy, it is important that you understand a little bit about how energy works within the human body. I'm not talking about "the energy to get up in the morning," although that can affect the sort of energy I am talking about, which is called your *energy body*.

YOUR ENERGY BODY

If you are reading this book, you are alive and living. You have the energy and ability to think thoughts, feel feelings, move around, and affect others with your thoughts and actions. You are a vibrant life force, and as a life force you have a measurable vibration—many of them, in fact. Your voice has a vibration when you speak, your breath when you breathe, your heart has its own rhythm of beating, your stomach massages its food, your brain pulses as you think.

Even when you are unaware of particular thoughts, your mind is always taking in stimuli, sorting them out, making your own particular sense out of them. Perhaps your mind is busy figuring out your schedule and—how will there ever be time to fit the laundry in? But you must! As you see an image of that shirt you want for Tuesday literally float across your mind's eye, you're still reading this. At the same time, your brain is also cleverly sending all sorts of important signals through what we call the autonomic nervous system which takes care of all the details of being, like growing hair and nails, repairing the little cut on your finger, and so on.

Vibrations can be considered actual musical tones. We can think of ourselves as creating a symphony of a sort from all the different vibrations we put out. It is easy, then, to see why we can be in harmony within ourselves. We could also consider all the different vibrations we have as being overtones of the *one* complex overall tone that we are.

You constantly exude these vibrations. They extend outside your physical body. These are emanations of your own spirit, or soul, or life force. This emanation of vibration from your physical body is real, and it is detectable. Kirilian photography has captured the proof on film. What it looks like on film is a "halo effect" around the outside of the physical body. This is often called an aura. The aura has been painted by master painters as halos around the heads of saints and angels. The truth is that we all have auras, and you can feel yours, even if you can't see it. It extends around your entire body.

Exercise 1: Energy Field Awareness I

If you are not already aware of your energy field, read the following exercise through, then proceed.

1. Put down the book and the crystal.
2. Get comfortable sitting on a chair or cross-legged on the floor.
3. Take a couple of easy breaths through your nose, and clear your mind.
4. Now close your eyes to limit the outside distractions of light and color.
5. Rub the palms of your hands together, and cover your face with your hands, but do not touch your face. Keep your hands about three inches away. It may take a few moments of being open to the experience to feel what takes place. You will have to just listen for a moment, but not with your ears. Feel free to experiment a little—move your hands back a little, then a little closer.

You probably experienced a number of things. One may be a feeling of heat, either from your hands or from your face. Or maybe it was a spongy sort of feeling of something being between your hands and face. Maybe you played with the distance and felt a pressure against your hands or face as you got closer again. All are possible experiences, and I'm sure there are countless others—it would be too difficult to enumerate them all.

This real but unseeable part of you is your energy body. If you played with the distances during the exercise, you may have noticed another level of spongy resistance outside the three-inch level, at about eighteen inches. If your hands could extend further, you might be able to feel another level of your energy body at about three feet away. Try the exercise again, if you like, keeping these other levels in mind. A few people can see energy bodies with their eyes, although it often takes a lot of practice to learn.

It might help for you to picture the outline of your body as a cookie-cutter shape, and your energy body as a series of progressively larger cookie shapes around you. Most holistic healers believe that there are at least three levels of body energy (at approximately three inches, eighteen inches and thirty-six inches). Many people, including myself, believe that there are others as well. Much as your internal vibrations can account for feeling "in tune with yourself," energy body vibrations can put you "in tune" with someone else.

Let's quickly run through one more exercise that will bring your energy body into your conscious awareness. Read it through, and then proceed.

Feel free to repeat any of these exercises as often as you wish.

Exercise 2: Energy Field Awareness II

1. Sit comfortably, take a few breaths, and clear your mind.
2. Rub the palms of your hands together lightly a few times to bring some extra energy to the area.
3. Separate your hands an inch or so. You should be feeling the first layer of your energy body at about one to three inches.
4. Slowly separate your hands, and see how far apart you can let them get and still feel "something."

This feeling can be detected anywhere along your body. The hands tend to be more sensitive in detecting the energy body, but that's how we are built, our hands being two of our most highly attuned sensors.

The energy body demonstration can also be done with two people. For optimum results, both people must be calm and pay attention. Using either hand, you can detect another person's energy body anywhere a few inches above his or her physical body. The person being touched may or may not feel any sensation. Here again, it is a matter of being sensitive to energy; there are no shoulds, rights, or wrongs.

Your energy body is part of you; it is several more layers of you than you thought you were. Like the you that you are more familiar with, your energy body can feel things, remember things. It can experience. It can get injured and can heal. It can get dirty—and you can clean it.

HEALTH FOR YOUR WHOLE SELF

Health is not merely a matter of absence of disease; your health is the sum of your parts. A healthy physical body gives you a healthy energy body. A healthy energy body gives you a healthy physical body. Although this sounds a little like a chicken-and-egg situation, let's just say for now that it's a simultaneous truth.

More and more, scientists and doctors are beginning to acknowledge what holistic healers have known all along—that emotions can affect our overall health. And emotions affect our energy bodies as well as our psyches. When someone says something that hurts you, it is first a thought. Thought is an amazingly strong vibration. Often someone need not say a word, and yet a message is loud and clear. This thought, which may or may not be articulated, has an electrical charge. So does your energy field. Your energy field absorbs all of these experiences, both good and bad. It is not judgmental—it stores everything.

Let's look at what can happen to you when you're feeling "good" versus feeling "bad". We as humans are very changeable, but don't confuse this with moody. We grow and change and adapt to what comes our way. This is our nature, and is understandable when you think of us as being made of mostly water. We are liquid, metaphorically and physically.

We're all familiar with the terms *good vibes* and *bad vibes*; we have all felt someone else giving us good vibes. A good friend perhaps, with whom we are reunited after a separation. He or she is all smiles and glowing, and we probably give good vibes back. As you picture this in your mind's eye, you are, in part, reexperiencing the feeling. You may be smiling and feeling the emotions of that moment past, only this time a little less intensely. Take a second and remember a pleasant feeling. It feels expansive and loving. You feel like all of you is reaching out.

Now I hate to blow your

good time, but find an experience that is a bad vibe. The person that was behind you in line at the supermarket—he was angry and pushy and topped it off with some expletive. Remember an incident like that. How does it feel? Contracting, gray. Not easy to shake.

When you are surrounded by love and in a supportive environment, when you are doing work you love, then every day's a holiday. Your energy field glitters with electrical sparks of love and approval. Chances are you're also in perfect health and life is grand. No one's life runs perfectly smoothly all the time, however. Sometimes things can be tough, and life can seem full of stress and even overwhelming. This is when it's a little harder to feel emotionally, mentally, and physically in top shape. When a good feeling permeates your energy field, it's good. When feeling bad gets stuck in your field, it's time to get rid of it. A constant flow of bad vibes detracts from your physical or emotional health.

Here's a hypothetical example: Pat works for Joan. Pat is very open and outgoing, Joan less so. Pat feels constantly overtaxed. There's

A word about the terms *positive* and *negative*. These terms, as I'm using them, have no relation to positive and negative electrical charges, or poles, or ions. These references are to the quality of the energy or vibration: Positive vibes elicit good feelings and are good for you; negative vibes have the opposite effect. These terms are descriptive and nonjudgmental; you alone must discriminate those things that are good and bad for you.

so much work to be done and so little time. Joan can be demanding from time to time, but she, too, has a lot of work to do. So, the pressure is on, and Pat is convinced that Joan is a pain in the neck because she has given Pat so much work to do. Joan is angry that Pat is always complaining that she has to work overtime. It's no mystery that over a short period of time, Pat's shoulders are up to her ears. Before you know it, she's living on aspirin to kill the pain in her head caused by neck tension. And Joan is even more angry now, because Pat is ragging on about her headache as well. You don't have to be Sherlock Holmes to figure out how Joan developed indigestion, also known as heartburn.

While this story is only one example of how negative thought can impair emotional and physical health, I'm sure you recognize a pattern that you, or someone you know, has fallen into before. Experiences such as these suggest three things:

1. Thoughts are very powerful, and unless negative ones are released, over time, they can lead to physical manifestations of pain and disease. These negatives can be picked up anywhere: on the street, in your neighborhood or at the store, from a co-worker or a dear friend who is "down." There is no reason to literally carry around someone else's unhappiness; it is important to dislodge negative thoughts from our energy bodies.
2. It's important to try to be around people who project "good" thoughts and attitudes.
3. It's important to develop "good" thoughts and attitudes within yourself, and to project them.

Now I realize this sounds a little didactic, so I should point out that we are always making decisions for ourselves. We must learn to make choices that are good for ourselves as individuals. If you know anger is good for you (though I doubt it), be angry. No one can tell you what thoughts are "good" for *you* to think.

Let's have a look at some of the negatives that most of us will want to avoid:

- **Anger**
- **Greed**
- **Worry**
- **Fear**
- **Obsession**
- **Envy, jealousy.**

All these create undesirable energy. It's also pretty much a tossup whether it's worse to be on the sending or receiving side of any of these negatives. Keeping free of negative energy is important to our health and well-being, for the short term and the long term.

Because energy seeks a balanced level, negative energies will be attracted to positive ones. Maintaining your good vibes against the daily onslaught of bad ones can be accomplished in three ways:

1 By cleansing the energy body

2 By replacing any negative thoughts with positive thoughts for happiness and growth

3 By eliminating the source of negative thoughts, either by physical removal or by overpowering negative thoughts with positive ones

We will learn ways to do this in the subsequent chapters.

To give yourself time to digest all of this, turn your attention to your quartz crystal.

Energy that expands is giving, opening, has the qualities of expression. Energy that contracts is drawing (of energy), closing, has the qualities of retention.

4 ♦ THE POWER OF QUARTZ CRYSTALS

Put aside any jokes about pet rocks at this point, and take a fresh look at the stone. It is so perfect and complete, with its millions of molecules in perfect alignment. It seems to have its own light.

This stone, like you, has its own natural vibration, its own energy body. It vibrates at an unvarying frequency. This is *un*like you, who are always undergoing change. Not only does the stone you are holding have a frequency, but it has an energy field similar to yours. And, like your energy field, a quartz's energy field is not obvious. It can usually be sensed more successfully than it can be seen.

As a quartz crystal vibrates, it conducts an energy that has positive results on the energy bodies of people. If there is any doubt in your mind that a seemingly inanimate object like a *rock* can actually affect your health, I'd like to ask you at this point to consider the repercussions of being very close to uranium. Uranium is a dangerous radioactive substance that looks like an innocent rock. There are no easy signs to tell you that it is vibrating at a rate that is completely destructive to your physical body. While uranium has been found to have a bad

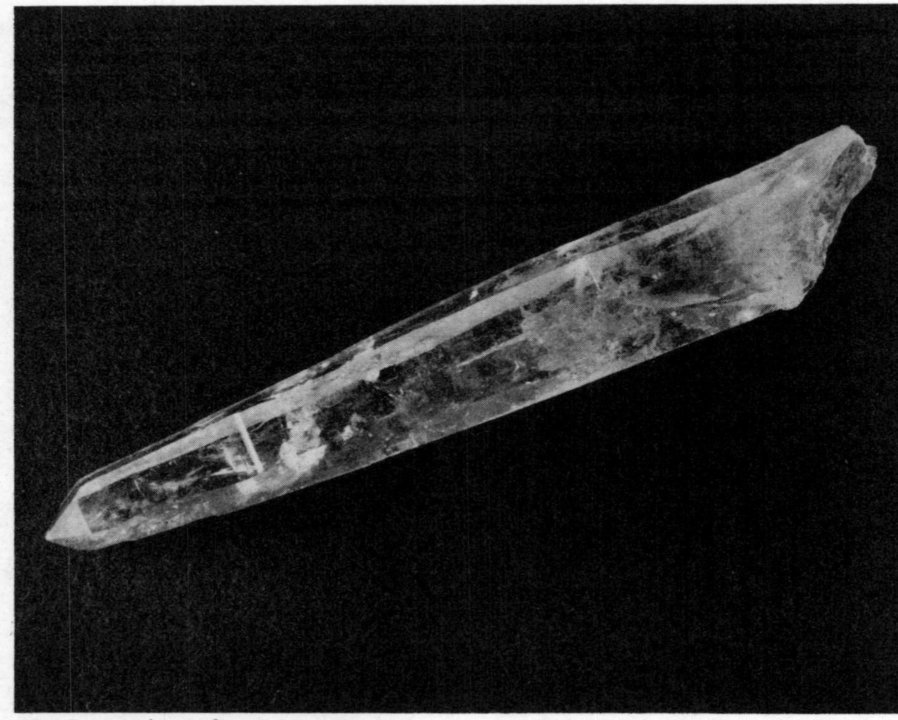
Single-terminated quartz crystal.

effect on us, quartz and other gemstones actually have a positive effect on us, and can be used in a number of ways to benefit our physical, emotional, and mental selves.

CRYSTAL'S HEALING ENERGY

The reason crystals are healing agents is simple and complex at the same time. The crystal seems to draw life breath, or life-giving force, out of the air, much as we do. In Hindu philosophy, this life force, or vital energy, is called *prana*, and includes all energy forces, including gravity, electricity, and magnetism.

Although the natural inclination of our bodies is toward health, life, and

The energy that flows through and around a crystal can be likened to the magnetic field that flows through and around a magnet. The pattern of energy flow in a single terminated crystal resembles that of a bar magnet.

balance, humans are, as you know, changeable by nature. Crystals, however, are not so changeable, and are constantly exuding a positive life force. A crystal in your environment can bring you back to a healthy vibration, because of the (liquid) body's tendency to come into harmony with the more constant, and harmonious, vibration of the crystal.

Remember, the physical and energy bodies are not separate. Each one affects the other. If your emotional body is not treated and stays constricted in an area, your physical body can get better but will not heal completely until the emotions are also healed. Or if your physical body does manage to heal completely, a relapse or a reinjury may occur until the energy body is healed. The bodies are then returned to a state of balance, and you should be sure that the state of balance is positive for you.

As an example, if Pat goes to a chiropractor for her neck pain, it would help the chiropractor, and Pat, if there were a crystal in the work room. While the chiropractor is working on the physical neck problem, the crystal is moving the negative energy block from the emotional portion of her energy field. It makes the job easier for the doctor, and Pat feels better faster.

Now let's say that Pat has fallen on the ice and broken her arm. Poor Pat. We will look at this injury as a physical injury, without getting into why and how she fell, which may actually have to do with energy blocks. It is still beneficial for Pat to use quartz. The additional positive life force will help her bone heal. This will also help heal the damage done to her energy field because, as you remember, the health of one body affects the health of the other. Getting the bone to heal quickly will have other good repercussions—she will lose a minimum of muscle tone, because she'll be out of the cast sooner.

Pat could apply quartz directly to her wounded arm. Let's say she tapes a crystal to her cast. It will help it heal faster, as mentioned before, and at such a close range, it will draw away a lot of the pain. It is important to help Pat feel more comfortable, and use of a crystal may help her to use pain relievers less often. This is good because the pain reliever, the crystal, is working the spot where the pain is felt, instead of on the whole body.

Your crystal will channel your vital energy so you have more of it at your disposal. When using a crystal as a tool, it is up to the user to channel this energy consciously, and in a really

In general, having a crystal means that you have more vital energy available. This can give your physical body more strength, and you may react from time to time by feeling rejuvenated, or just less tired. One word of caution: You should never overtax your body. Although it does not necessarily have this potential in all people, don't be tempted to use your crystal to keep yourself energized to the point of sacrificing sleep. Sleep is important in your life; a tremendous amount of healing takes place in the sleep state, and a lot of mental sorting out takes place, too.

positive direction. There are a number of specific ways to do this, which we will go into after covering a few more basic concepts and preparation exercises.

You've already experimented with finding your own energy body. Let's see if you can become sensitive to the energy body of your crystal. Read Exercise 3 first, then put the book down and try it.

Exercise 3: Crystal Vibration Awareness I

1. Find a comfortable sitting position. Relax, and try not to anticipate any particular result.
2. Hold the crystal lightly between the fingers of your right hand (if you're a lefty, use your left hand). Relax. Close your eyes and take a few easy breaths to help you focus your attention inward.
3. Open your eyes, and, holding the crystal, aim the termination at your left palm, about an inch away.
4. Slowly, draw a circle with the crystal, never actually touching your skin. Give yourself over to the exercise. Continue drawing circles. If you have no particular sensation, make sure you are giving yourself enough time to tune in. If, after a full sixty seconds of attention, you feel nothing, do not be discouraged. It may take some time to become sensitive to these subtle energies.

Quartz works just the same whether you believe in it or not.

You may have felt nothing, or you may have experienced any number of sensations. Perhaps you felt a tickling on the palm or a sense of pressure. Maybe your fingertips felt more blood rushing to them. You could have noticed a number of sensations. Maybe none. This is not a test; whatever you experience is correct.

Let's try one more sensitivity exercise.

Exercise 4: Crystal Vibration Awareness II

1. Place the crystal in your left hand if you're a rightie, right if you're a leftie. Hold the crystal gently in your receiving hand.
2. Go through the previous exercise's routine: Sit, relax, breathe, clear your thoughts to focus your attention on your hand, and just listen, see, and feel.

You may have felt nothing, or you may have felt something—a swirling but not unpleasant feeling in the pit of your stomach, across the brow, or around your eye area. You may have felt a tingling sensation in your palm or up your arm, or perhaps a slightly numb feeling. The crystal may have rested in your palm and felt heavy then light in a rhythmic pattern. Did you hear a tone, not exactly with your ears, but sort of in your head? Perhaps you felt a sensation of heat in your palm.

If you felt nothing, please be patient. Don't throw in the towel just yet. This type of sensitivity is not always available the first few tries. You might try to limit your distractions. Go into a room by yourself if necessary. Spend more time preparing yourself. Also, try not to anticipate any particular reaction. Everyone has different reactions to stimuli. You may not "sense" exactly the way I do or your best friend does. *Trust your own experience.*

USING THE CRYSTAL

There are two main ways to use quartz, *passively* and *actively*. Just having crystals in one's environment can improve the quality of life, but they can also be used in personal or group meditation, or in conjunction with other healing techniques. Crystals are being used in almost as many different ways as there are different healers using them.

Passive Use

Using crystals passively is pretty self-explanatory. You keep a crystal around—in your workroom, bedroom, kitchen, bathroom—anywhere you might want to increase the energy force. Pat used a crystal passively when she

You have a sending hand and a receiving hand. If you're a rightie, your left hand is probably your more sensitive hand. Also, it is electrically on your negative, or receiving, side, while your right hand is on the positively charged, giving side. Lefties are probably reversed in their electrical spiral, and may use the right hand for receiving. Not all lefties, however, are mirror opposites of righties, and both lefties and righties can tend toward ambidexterity. The key here—and throughout the book—is to *do what you feel is correct for you.*

With crystals, bigger isn't necessarily better. All sorts of things combine to determine the power and vibrational frequency of a crystal, including size, clarity, and where it was formed. Generally (and this is *very* general), a larger crystal will have a greater radius when used passively—its power will travel farther—but a less intense vibration than a smaller stone might have.

taped it to her cast.

The crystal that came with this book has an effective radius of about three feet. In general, the radius of the energy field of a crystal increases in proportion to the size of the crystal. A room can be filled with quartz energy by placing small crystals at strategic points in a room. You might want to put a crystal in each of the corners, with points all facing inward, or with points aimed at the back end of the next crystal in sequence, following along counter-clockwise.

A cluster of quartz can be effective, too, as the crystals seem to amplify each other. In other words, two small crystals placed together tend to act synergistically; they can conduct an amount of vital force greater than the sum of the energy they could conduct apart.

Another method of passive use is to place a large crystal, often called a generator, someplace in a room. Putting a crystal near a window where it can be charged with direct or indirect sunlight is ideal. Orienting the crystal in a north-south direction seems to work best. The termination point should face to the south.

The ambient use of crystal can be beneficial in a work environment because of the way a crystal seems to deflect negative energy and calm the nerves while energizing the system. Another place that a crystal can serve an important function is the workroom of a healer, as we saw with Pat's chiropractor.

For any healer—a specialist of holistic medicine, a massage therapist, a psychiatrist, a dentist, or a witch doctor—the subtle energy that flows through the crystal is beneficial and healing. The healer often finds that removing a block from the psyche or easing disease from the physical body can be done more efficiently and effectively when the flow of energy that surrounds a crystal is present, helping the patient (usually without his or her realizing it) to get rid of some of the tensions, or contractions, that may have caused the problem in the first place.

The last and probably most obvious passive use of crystals is to carry one in your pocket, or wear one as jewelry. Be sure to put your stone in a pouch or in a bit of cloth to protect the delicate point and edges from being bumped by change or keys.

Active Use

Moving a block from an energy body may require more than the ambient use of crystals. This is where the active use of crystals comes in—for direct healing, meditation, and cleansing. The subsequent chapters of this book teach the fundamentals of self-cleansing and meditation (focusing) with your crystal, and touch upon the art of self-healing with these amazing stones.

Removing blocks from the energy body can bring up into

It is important to stress that the subtle energy a crystal imparts is always present in the atmosphere. The crystal takes already existing energy, focuses it, amplifies it, and then directs it. Although the crystal has a field of energy all around it, the exit point of directed, amplified energy is at the crystal's termination.

a conscious level emotions that may be challenging to resolve. Assume, for example, that Joan has been meditating with her crystal. "Why," she is asking herself, "do I get the pains of heartburn on the evenings Pat and I work late? Because I don't enjoy my work enough?" No, no, she thinks to herself, because there is no affirmation in her heart when she offers the suggestion. She offers every idea she can think of. No positive response. So Joan just sits with the crystal for a few more moments and allows herself to drift, keeping an open mind and heart, listening for the inner voice to answer. Then suddenly she realizes: "It's Pat's complaining! I can't stand it, and I have trouble telling her, because I certainly don't ever want to tell someone what to do." Joan finds in her heart the affirmation that she needs. *This* is the basis of her heartburn.

She has discovered her block. Now she is consciously aware of the problem, which gives her an opportunity to remove the block for good. To do that, Joan must find the appropriate way to deal with Pat, but it is possible that the heartburn block is the result of a deeper issue within Joan. She may have to overcome the more complicated problem of expressing herself to others, especially her discomfort with telling people when they are stepping over the line with her and irritating her. So Joan has to learn to better recognize when she is becoming irritated with good cause, and then to better express what is in her heart.

Joan can use her crystal to approach this problem in a number of different ways.

- *She can* program *her crystal to give herself greater sensitivity to her own feelings and to give herself courage in facing others. (For instructions, see section on programming.)*
- *She can sit quietly, and hold the crystal over her heart area, the termination point facing out, to open her heart energy center.*
- *She can clean her energy field every day or every other day to help dislodge negatives that may be*

clinging. (For instructions, see section on cleansing.)

- *She can create a pattern of many crystals on the floor or on her bed and lie in the pattern for fifteen minutes. This is cleansing and rejuvenating, and an excellent time for meditating on removing her block and using affirmative thoughts to change old, outmoded thought patterns. (See the illustration of Joan in a pattern. She is using six rose quartz chunks and two clear quartz terminated pieces. Please note the direction of the terminations; they are both pointing the same way and directing the flow of energy from bottom to top.)*

Joan in a crystal pattern.

- *She may choose an appropriate crystal pendant and wear it on a chain that places the crystal over her heart.*
- *She may want to seek or create some sort of group of people interested in the use of crystals. Perhaps she would choose a group whose goals are directly related to the particular block she is trying to remove. For example, if you are trying to remove the habit of smoking cigarettes, you may want to join a group formed just for the purpose of stopping smoking. I quit cigarettes after twenty years of obsessive chain smoking with the help of crystals. But as a friend of mine says, "Whatever floats your boat." Everyone has a unique approach to reaching goals.*

You will know your correct path when you open your heart to yourself and give yourself the love and courage to implement what you know is right for you.

5 ♦ CLEARING AND CHARGING YOUR CRYSTAL

A crystal can pick up and temporarily hold a charge. As you pick up charges from the environment or experience, so can a crystal. However, crystals are generally easier to clear of vibrations and are *infinitely* more stable than we are. If you do a cleaning with your stone, the stone acts like a clothes brush you use to whisk lint off your jacket. You must at some point remove the lint from the brush so that it can continue to do its job. When you use a crystal, a similar principle is at work.

Therefore, before you cleanse yourself with a crystal or use a crystal in any active or passive manner, you might want to make sure the crystal is clean, or clear of any disruptive vibrations. Determining the state of a crystal is fairly subjective and intuitive. If it seems clear, then it is, for you.

WHEN TO CLEAR AND CHARGE

Every so often you might want to clear the crystal's energy field and charge it with new, "fresh" energy. When you charge the crystal, you deliberately place or replace in the crystal a pure or impartial health-giving vibration, such as the energy of the sun and the ocean. Charging and programming (which we'll discuss in Chapter 8) are similar: Charging is energizing; programming is charging with a particular intent.

You might want to clear and charge your crystal after you've been ill, or you might want to clear and charge it when you've just purchased or received it. Clearing and charging your crystal can be an opportune time to familiarize yourself with your crystal before you actually start to work with it.

Just as there are all sorts of reasons to clear a crystal, there are all sorts of reasons to leave it alone. You might not want to clear your crystal right away if it was a gift of love from a friend and has been cleared and programmed with you in mind. Or you may not want to clear a stone that you've purchased because there is a "certain something" about it you think will be lost. As you become more familiar with the stones and their use, you will find what works best for you.

HOW TO CLEAR AND CHARGE

There are many ways to clear and charge a crystal. Feel free to make up variations on the list that follows; do whatever feels right for you. All the following methods work well.

Method 1: Cleansing in Warm Water

Cleanse the crystal in warm water with mild soap and a sponge or a soft brush (an old toothbrush works well). There is no substitute for a good physical cleaning to make your crystal sparkle. This removes any dirt or oil that can make your crystal sticky to physical and nonphysical goop.

Method 2: Cool Tap Water

After following method 1, or as a separate clearing, you can run cool tap water over your stone for at least thirty seconds. This rush of cold water seems to clear and charge the stone.

Method 3: Naturally Running Water

It would be ideal to place the stone in a running stream or

in the ocean for cleaning, although these are not always available or convenient. If you can take a trip to any of these natural waters, all the better. To wash your stone in a stream, place it in a fabric bag or a mesh basket and tie it to something stable. Let the stone sit in the rushing water for as long as you like. A half-hour is a good length of time, but five minutes would be sufficient.

This method of washing with natural water can be done at the beach. Gather your crystals and take them for a swim with you in the ocean. Or just give them a quick rinse in the surf. The natural salt content in ocean water cleanses crystals quickly and thoroughly. On these excursions, you need not pack extra beer and chips for your mineral friends.

Method 4: Solution of Sea Salt

If you'd like to wash your crystal in salt water, but cloudy days, winter temperatures, or distance prevent a picnic by the ocean or lakeside, you can clean your crystal at home in sea salt—crystalized sea water.

Make a strong solution of sea salt (kosher or coarse salt is fine) and cold water in a glass or ceramic container. Do *not* use metal; the metal and salt may react and just complicate matters. Soak your crystal anywhere from a few hours to overnight. When you feel your stone has soaked in the salt water long enough, give it a thirty-second shot of cool tap water.

Note: This process is *not* recommended for jewelry. The sea salt tends to react with metal and can cause discoloration or even pitting, depending on the strength of solution and the type of metal. Also, keep the bowl with your stones and the sea salt out of the direct sunlight if you are clearing any *colored* stones, as you run the risk of bleaching the color out of the stones.

Method 5: Direct Sunlight

After any of these methods of clearing, you will be ready to further clear and charge your crystal. Place the stone on a clean white surface in direct sunlight. If you have only indirect sunlight, that is fine. In a pinch, you can put the crystal by an open window. Placing the stone in sunlight will charge it up with the energy of the sun. This energizes the crystal with a particular type of energy that is active and mental by nature.

You may choose instead to charge your crystal with moon energy, which is more intuitive and emotional by nature. If you leave your crystals by the window for long periods of time, chances are good that both types of energy have been infused into your crystal. One is not better than the other; each has its own individual attributes.

Be advised that some colored stones can be bleached by too much exposure to direct sunlight. I've been told that the purple in amethyst is particularly volatile. I have never had any mishaps with amethyst myself, but I feel it's necessary to pass on a word of caution, so keep your eyes peeled.

Clearing your crystal and charging it can often be accomplished simultaneously. For example, exposure to the sun will both clear a crystal of old energy and recharge it with new. Sometimes, however, old energy won't be gotten rid of simply by exposure to new, positive energy; while the crystal is charged with new energy, it also retains the old. I usually do a salt-water cleansing before charging in the sun, but it is not always necessary. Do whatever you feel your crystal needs. If you desire a quick cleanse and charge, cleansing and charging with intent is particularly effective.

Clearing = Cleansing = Cleaning

Method 6: Cleansing Crystal with Crystal

Another cleaning option is to place your crystal on a larger crystal or on a crystal cluster. You could also arrange a few large crystals in a pattern with termination points facing toward the crystal you want to clean. Four crystals, for example, could be used in a plus-sign formation, with the crystal(s) to be cleaned placed in the center. Three crystals could be used also. Be creative. Any pattern you think will work, will. I like to set these patterns up in sunlight, but they work equally well out of the sun.

Method 7: Fragrant Dried Flowers

A personal favorite of mine is to place my jewelry or a favorite crystal in a dish of dried flowers mixed for their fragrance. I keep it near my bed so that I can have my favorite jewels near me while they are cleaned overnight by the dried potpourri. I also get to smell the lovely fragrance. Amazingly enough, the stones retain the scent for a few hours. You only need to change the flowers every few months.

Method 8: Cleansing with Intention

Cleansing your crystal with intention is effective and requires a minimum of paraphernalia. Hold the crystal in your giving, or sending, hand (right for righties), and *will* your crystal to be clean. Allow your thought pattern to override any pattern that has been introduced to the crystal. Breathe consciously during this process. With every inhalation, gather strength to your intention; on your exhalation, "blow" unwanted energies away "through" your hand and out of your crystal. You could also "blow" these energies away from your heart center or from the center that is located on your forehead, often called the third eye.

This process is almost identical to "programming." If you feel comfortable cleaning your crystal with intention, you may want to program it to be an "automatically clean" crystal. Or you might want to do this after a more physical cleaning. To put your crystal on automatic clean, place your crystal in your sending hand and *will* (program) your crystal to deflect and not absorb energies that you yourself don't place there. This should cut down on the frequency with which you must clean your crystal. It should also make you feel easier about having someone who isn't your favorite person touch a crystal you may be carrying or wearing.

If you feel strongly about not letting people touch certain of your crystals, this is understandable. Keep that special pendant inside your shirt or have that crystal tucked away in your pocket. You may want to have a crystal or two that you can share with others.

Method 9: Clearing with Sage

Another fragrant clearing method involves sage, used by chefs and, traditionally, by the American Indians in rituals. When burned, it has a clearing effect in the air, on our energy fields, and on a crystal's energy field.

Buy your sage in a crystal or other specialty shop. The sage you find at the supermarket in jars and cans is the same plant you can find in the shops, but the big difference is usually in price; the sage meant for eating may be more carefully selected for flavor, and is usually crumbled into tiny bits, which, for your purposes, makes it difficult to handle. (You might also find "smudge" sticks, often combining sage with sweetgrass and cedar. Or you could use sweetgrass alone.)

Let's say you have bought some sage or you know where to pick it wild, and you have dried the leaves. Take a large sea or abalone shell (traditionally used) or metal bowl, place the sage in it. Light the sage so it catches the flame, and then put it out. Using some sort of cover for this is helpful and safe. The idea is to get the sage to smoke, not necessarily to burn up. Now, run your crystal through the *smoke*, turning the crystal to get all sides. Do this till you feel intuitively that the crystal is sufficiently cleansed. If the sage is still burning, you may want to use it to clear the room. To clear the room, carry the sage to every corner of the room and let the smoke drift everywhere. This can be done at any time—before or after a meditation or healing can be nice.

The words of caution about saging are pretty obvious. Always be careful with fire. Do not use much sage—if it all catches fire, you want to be able to handle the flame. Have some water handy the first few times you try this, until you get the knack of it. If you decide that the smudge sticks appeal to you, make sure that when you put out the stick it is *really* out. Dip the end in water. The stick will probably be dry by the next time you want to use it, and this shouldn't damage the stick.

Method 10: Tape Demagnetizer

Use a tape demagnetizer. Simply pass the demagnetizer over the crystal, following the directions on the package. There is also a demagnetizer used for televisions. This is OK, too.

A word here about electronics: Don't place your crystal directly on electronic equipment when you are charging them or displaying them. TVs, stereos, radios, and microwave ovens are constantly putting out massive amounts of current that can affect the crystals, giving them an overload of electrical energy. The exceptions may be placing a crystal on an appliance to help the appliance be more energy-efficient, or if you work at a computer terminal and wish to have crystals with you.

DEEP CLEANSING

The following methods of clearing aren't particularly quick, but they are effective when you want deep cleansing.

Deep-Cleansing Method 1

Bury your crystal in sea salt. Place sea salt in a noncorrosive (nonmetal) container, and bury your crystal in the salt. It is best to let the stone sit undisturbed for three days and nights. If you feel that the stone needs an especially intense cleaning, you can leave the stone for seven days and nights.

After completing the clearing, you can charge the crystal to revitalize it further, with the sun, moon, or air. Toss the salt in the garbage when you're through with the cleaning—never use the salt on your food after it has been used for a clearing.

Deep-Cleansing Method 2

Bury your crystal in earth. Use a similar time frame as in the salt method. Three days should do it; seven days is bringing out the big guns. In dire straits, bury the crystal on a new moon, and dig it up when the moon cycle returns to a new moon.

Don't place a crystal to be cleaned in your favorite house plant. There is just not enough dirt in the pot to clean the crystal and modify the energy you are placing in the pot without having it affect the plant. When cleaning by the earth method, use a plantless pot of soil or clay, or bury the crystal in your back yard. The earth can handle the energy that we consider negative and convert it into useful energy. If you bury the crystal, don't forget to mark the spot well. You only *think* you can remember where you put it. After you retrieve it, rinse it with water, and you are ready for programming or charging.

At times, although your crystal has been cleared and you have charged it with sun or moon energy, it may seem to be lifeless or sleeping. There doesn't seem to be a good reason for why this is happening. It could mean that the crystal needs an extended period of rest. This is especially likely to happen to jewelry that you wear daily. This could also have to do with the planetary influences; crystals have always been attributed with astrological connections.

There is also the possibility that, although the stones do not change, our receptivity on a conscious level does. As I've mentioned, we humans are very changeable creatures. We go through all sorts of cycles, often referred to as biorhythms.

Generally, leaving a stone alone for an extended period of time will allow it to clean and recharge itself spontaneously.

Let your intuitive self choose the method you feel to be the best for you, and for each individual stone.

6 ♦ CLEANSING AND GROUNDING YOURSELF

Let's focus now on cleansing ourselves. You have cleaned your crystal, and through learning how to clean a crystal, you have actually learned a great deal about how to clean yourself. But whatever you do, do not bury yourself alive in sea salt.

In all seriousness, we are too often stuck working in environments that give us little or no natural sunlight. Too often we are indoors and breathing recycled and warmed or cooled air. When we finally get a chance for a vacation, chances are we want to get out there and play too hard, too. With so little time for recreation, we try to make the most of it, often neglecting the time we need for quality rest. Sometimes we become ill just because we need some time off—it is our body's way of telling us to slow down.

Sometimes we need physical rest, and sometimes we need mind rest. There are other reasons to need rest. Perhaps there have been many changes in your life, changes in living conditions, changes in major relationships. Perhaps you need time out to digest and integrate these changes before you can go on with your daily life. Rest and sleep are some of the greatest healing tools we have.

Meditation, and some of the work you'll be learning here, can help tap into a state that in some ways is similar to the sleep state, thereby accessing great healing potential.

CLEANSING TECHNIQUES

There are several ways to cleanse yourself. As with crystal cleansing, choose whatever appeals to you.

Physical cleaning is one of the best ways to soothe the nerves or "rinse the day off." After a long hard day, a soak in the tub is relaxing and rejuvenating, and a wonderful thing to do before starting some of the other less physical cleanings and meditation. Mineral salts in the tub can be extra soothing; so can bathing with your quartz or other stones. Covering up with sand when you're at the beach is a favorite of many, and is a great way to clean your energy field. After all, the beach is primarily quartz: All those little bits of sand are grains of quartz, being cleared and charged by the salt air and the sun daily. The most obvious, and too often forgotten method of cleaning, recharging, and basically getting healthy, is being sure to get rest, fresh air, and sunshine.

Music is a tremendous healer when you are listening to the right music. Music can lift you to the heavens, or it can be destructive; selectivity is of great importance. The music and rattles and drums of witch doctors are based on the great effect of the vibrations of sound. Choose your music to inspire your mood.

The highly developed science of mantra of the Eastern religions can resemble singing when it uses the voice, but mantra can also be done in silence, solely in the realm of thought. Briefly, mantra is the repetition of Sanskrit syllables that make up words that describe and praise different aspects of heavenly spirit as perceived on earth. There are also mantras of an abstract nature.

The syllables themselves have the ability to shake up and wake up and clean different centers of our physical and energy bodies. These energy centers, or "chakras," are becoming a part of our vocabulary and our experience in the West, and are important in the work that is done with crystals. We will run through these concepts in detail in another section. Mantra is very powerful and should be learned from a teacher who is trained in mantra and has gained and earned your trust. In this and in all aspects of your life, you should maintain a note of discrimination and control over your circumstances and the input you receive.

Prayer is the closest that Western organized religion comes to mantra. The major difference is that prayer is effective directly as a result of the thoughts or ideas that are being processed through the prayer, while mantra's effects are based directly on the exact vibration that is set up through the specific sound. A specific prayer to Jesus may be said in varying word arrangements and can be said in different languages and is as effective in each language. Mantra is done in Sanskrit only, and there is great attention to correct pronunciation. (*OM* is pronounced by a French devotee in exactly the same manner as by an Indian.)

Prayer and mantra are extremely effective cleansing tools. They raise the mind to thoughts or vibrations that are of a spiritual nature and are "higher" vibrations than thinking about mundane chores. Prayer and mantra are also helpful in cleansing oneself because the mind is busied with positive thoughts, and unable to worry about the clutter of negative thoughts of doubt and worry we all suffer with too often.

To summarize, here are some tips for cleansing yourself:

- *Try bathing, sand, fresh air and sunshine, music, prayer, or, as they say, "whatever works for you."*
- *You might try to get natural spectrum lights*

installed at home or at the office (see section on color and chakras).

- *Try to get some fresh air and natural sunlight at least once a day. Fifteen minutes could make a big difference!*
- *Try to get rest and plenty of sleep. Recreation can certainly be considered rest.*

All these can be extraordinarily important, as the lack of proper light, air, and rest have been found to be at least in part to blame for many of our ills.

You might also want to try the following exercise for calming the mind. This exercise is very simple, although it seems difficult at first. Give it a try—you'll find

Exercise 5: Stilling the Mind, or Focusing

1. Put aside five to ten minutes. That's all you need. You can increase the time of the exercise as you feel ready for it. Eventually, a concentration time of thirty minutes would be a good goal to aim for. (It's a good idea to start with two minutes or three, and have a timer, if possible, so you aren't distracted by wanting to look at your watch.)
2. Go someplace where you can be master of the space and not be subjected to distractions. If you must, try the closet, or I'm sure you could get a roommate to take a ten-minute walk (*and* take the dog). Affirm to yourself that the time will be beneficially spent on the exercise and only on the exercise. All other commitments will wait until the specified time is up. (Don't forget to turn the phone machine on, and volume *down!*
3. Sit in a comfortable position, but *do* be aware of your posture. Please don't slouch. If you need it, please place a pillow behind you to help maintain the curve in your lower back. Make yourself comfortable, loosen any tight belts,

continued on next page

continued from previous page

ties, or laces. If anything in the vicinity is bothering you, remove it, or find a less distracting place to do this exercise.

4. Close your eyes. Bring your attention to your breath. Keep your attention there, breathing fully. Keep your attention fixed on your breath. Feel the air come in your nostrils; it is cool air. Keep your attention fixed on the air as it leaves your nostrils; the breath has been warmed. Keep your attention on your breath. When your mind wanders, bring it back to your breath. A grocery list will appear; pay no attention. The breath is the assignment. An anger from last week will surface. It's not important now. That assignment from work . . . a moment to think . . . no, don't be tempted, bring your attention to your breath. Don't be angry or frustrated that your attention may wander. Just bring your mind back to your breath. Do not judge yourself during the exercise; do not judge the little cul de sacs of the mind. When you catch your mind wandering, gently

continued on next page

it will give you back ten times what you put in. It will allow you to tune in to vibrations around you with better clarity because you won't have the constant wanderings of your own mind grabbing at your attention. You may find yourself calmer after doing the exercise, especially if you have a nervous tendency, and you will find yourself energized if you are a little sluggish. In other words, this exercise will help balance you.

continued from previous page

push it back on the track till the timer rings. Don't be concerned if you hardly concentrated on your breath at all. There is nothing here to judge, nothing to win. This is not a competition. Not even with yourself.

5. When you have completed the exercise, take a little extra time to bring yourself back to the moment. Stay seated and open your eyes. Perhaps you feel like stretching; go ahead. You will probably feel as if you've had a nap and are ready to go ahead with your tasks. You might even feel relieved that the time is over. You might feel proud because you finished a tough exercise. You should feel much calmer and more relaxed than before you started, and also energized.

GROUNDING AND PROTECTING YOURSELF

Grounding is a descriptive word that means centering, or operating from your center. This is a state of being acutely aware of or paying attention to one's inner reactions, while remaining aware of one's outer environment. It can mean a state of being level-headed and in control of oneself, not easily knocked off balance by outward circumstances.

Protecting yourself by grounding helps keep your energy field less attractive to undesirable lower energies. As mentioned earlier, energy seeks a balanced level. If you have a strong, positive energy force and your neighbor a negative one, you can bring her "up," and she can bring you "down." This exercise is particularly good if you live in a dense urban environment, or if you work in an environment where there are hostile energies or where you feel vulnerable for any reason. You may choose to ground and protect yourself when you are experiencing unusual events in your life and would like to feel more secure. Many women choose to ground and protect themselves at particular times during their menstrual cycle when they feel most "open." You may find that grounding and protecting yourself will make a big difference in the way you feel at the end of the day.

A greater sense of self-esteem is a key factor to increasing the strength of your personal energy field as a deflective device.

There are many ways to increase the strength of your personal energy field as a deflective device. A greater sense of self-esteem is a key factor. Also, various yogic breath exercises are designed especially for circulating and storing prana, which is life force. One uncomplicated breathing exercise is simply to remember to breathe deeply a few times throughout the day. It's surprising how many people become engrossed in their daily tasks and forget to breathe properly. Proper breathing, by the way, is from the diaphragm, not just from the upper chest, and is always done with the mouth closed and through the nose. These are invaluable habits to cultivate if they are not yours already. You will have a chance to practice this type of breathing in the exercise that follows.

During this exercise and any other exercises that are suggested for your practice, you can choose to use your crystal or not. Think of your crystal as a tool; any self-healing or meditation you do is only expedited with the crystal. Do whatever makes you feel most comfortable. You may want to read through the book first before trying any exercise with a crystal. This is fine. You might want to try an exercise like the next one without your crystal first, and then try it again with a crystal.

If you want to use your crystal, hold it in your hand. Put it in your receiving hand if you feel that you will be receiving protection from without; use your giving hand if you feel that you are creating protection through your own will. Or you can simply put your crystal in your pocket. You might want to place it on a nearby tabletop or prop it up, point to the ceiling, in a shallow bowl filled with grains, rice, pebbles, or sand.

Whenever you are in need of grounding or energy during the day, feel free to tap into earth or heaven, pulling down energy from the sky when you are in need of a lift, or venting excess negative or frenetic energy as you need through the base of your spine, the bottoms of your feet or hands, or even

Increasing the amount of vital energy in your own energy body can help you handle or deflect the negative or unpleasant vibrations that come your way. Carrying a crystal in your pocket or wearing one as jewelry can help you increase your field. This seems to be most effective when the crystal (or faceted stone or shaped and polished stone) is carried often.

Exercise 6: Grounding and Protecting

1. Start this exercise sitting with your eyes closed. Breathe regularly, yet fully. Sit without leaning against a chair back if possible, and be aware of your spine. Breathe fully from the bottom of your belly, expanding your lungs completely but with ease. You may raise your shoulders gently to further expand your lungs, and gently lower them as you exhale, also fully and completely, from the top of your lungs to the bottom of your belly. Your abdomen should be expanding noticeably; don't be shy about that. Be sure that your mouth is gently closed and all breath passes in and out through your nostrils.

   As you breathe, be aware of your spine. Feel your "sit" bones on the seat of the chair, and feel the length of your spine as you bring breath to every vertebra. Allow yourself to gently sway with the breath, rising slightly with the inhalation and sinking slightly with the exhalation.
2. Imagine your spine extending downward as an imaginary thread. Imagine a heavy cord or rope extending from your spine downward, through your chair, through the floor, through the floors below, including the basement. Imagine the cord traveling through the layers of the earth, first a reddish layer, then a grayish layer, through rock, and finally rooted firmly into the core of the earth. You are now solidly fastened to the earth. Not even a tornado could budge you if you didn't want to move.
3. Now, leaving the cord solidly implanted in the earth, send your breath along your spine from the base, through your heart center and through the top of your head, up through the ceiling

and the roof, connecting with the god of your own heart. Imagine your breath coming out of your head as a slender and strong thread made of gold or silver.

4. Breathe, and with every exhalation, feel the earth and heaven connection becoming stronger and stronger.
5. Envision your energy field. Imagine that from every surface of your physical body there is a glow of life. This glow is particularly strong in the body area, along your lower abdomen, solar plexus, chest, throat, head, and along your back, the lower back, the middle back, and the upper back. See this as a protective orb around you. Imagine white, silvery light generating upward off the top of your head, like a crown. Be aware of your breath, as you slowly breathe in and out, calmly and completely. Every inhalation and exhalation increases your vital energy, like fanning a fire. You are feeling relaxed and rejuvenated. When you feel sufficiently energized, you may prepare to conclude the exercise.
6. Adjust the flow of vital energy: If you have any unneeded or unwanted energy, you can send it away merely by imagining it leaving through your cord to the earth. Or you can send it through the soles of your feet or through your fingertips if you prefer. (You may do this at any time you feel you want "vents" for excess energy, not just during this exercise.) Any energy needed at this time can still be supplied through the breath. You might want to get extra energy by breathing "through" the navel/solar plexus or from the top of the head. Feel free to tap into this energy source any time you need a lift.

continued on next page

continued from previous page

7. In conclusion, imagine a smoked glass covering over your entire energy body. Imagine yourself as a sunglass-covered M&M candy. You are now filled with vital energy, happy and healthy and protected from uninvited energies. If you feel the sunglass protection is unnecessary, then don't put it up. Follow your intuition. You may want to put a joyous blue tint to your energy field if you are feeling a little low.

your elbows. The earth by its nature can transmute any negative energy and use it to its benefit. Heaven has an unlimited supply of energy to feed you, so feel free—take as much as you need. You only need to tap in and take it.

Quickie Ground and Protect

Pay attention; this will really be quick.

1. **Send a grounding cord deep into the earth.**
2. **Send a silver or gold thread from the top of your head; tap into the sky.**
3. **Feel your energy body glow with vitality and pure white light. Imagine that your energy body places your physical body in an orb of protective light.**
4. **Put up your protective smoked-glass covering, or introduce a color. (Optional.)**
5. **You are ready to do a meditation or healing, or just to face the world.**

7 ♦ CRYSTALS, COLOR, AND CHAKRAS

When you meditate or perform self-healing exercises with your crystal, you'll find it immensely helpful to know what and where your main energy centers are. This information will also give you the knowledge that will enable you to use colored stones in your healing techniques. You will also have the option of programming your clear quartz with color vibrations to make your quartz function as a colored stone. If, for example, you desire the green healing ray of emerald, it is possible to program quartz with "green" because quartz is a gem of all colors. It is a clear stone, transmitting white light, and thus transmitting the full spectrum of color.

Knowing the energy centers can really help you gain insight into yourself. These energy centers were diagrammed by the Indians, in slightly varying ways, a few thousand years ago. It is no coincidence that these energy centers, or chakras, correspond to important glands and organs in the body. The word *chakra* is a Sanskrit word that is gaining acceptance as English vocabulary. This acceptance stems in part from the wide acceptance of this simple yet correct energy center "map"

Color surrounds us. It is present as long as we have light. It is so much a part of our existence that it is often not in the forefront of our consciousness. But it is always affecting us in subtle ways. It can make us sad or happy, speed up or slow down body functions, make us sleepy, or give us a feeling of being refreshed. Color can soothe us or turn us into frenzied maniacs.

among members of the crystal community, such as it is, and other New Age healers.

The sanskrit word *chakra* means, literally, *point* or *circle*. So while a chakra may emanate from a central point, it includes a spherical area around that point. Remember that when we say "chakra point," we also mean chakra circle.

CHAKRAS AND COLOR

Each chakra point has its own attributes or abilities. We usually think of perception and memory as being related to the brain only. Let us stretch our intellects now and consider giving the rest of our body an equal share in perception and memory. It is easiest to do if we relate each chakra point to its corresponding color. Color's effect on us is so intense that whole therapies have been based on nothing else. Our discussion of color, however, will be limited to its uses in crystal healings.

Color, like sound, vibrates. You may be aware of this vibration when you look at colors like "electric blue" or "fluorescent orange," but most of the time we think of color as being static. It might be easier to think of color as energy if you remember, from eighth-grade physics, what happens to a "clear" beam of light as it is passed through a prism: It literally separates into a rainbow of color. Think of these colors as light wavelengths. These wavelengths can be measured in the same way that sound wavelengths can be measured. But color is at a higher vibrational level than sound, so we can't hear it. At the low end of the color vibrational spectrum is red, which has the longest wavelengths. If we were able to modify our ears to "tune in" to color vibrations aurally, red is the lowest tone we would hear. Violet, at the top of the spectrum, has the shortest wavelengths, and is the highest tone we would hear.

Often, a person who is blind develops other senses to give him information about the world. A heightened sensitivity of touch is often developed, and some people can feel the difference between a red garment and a blue one. It would not surprise me if even a very sensitive sighted person could do this. You see, the red garment looks red because of the way it reacts to light; it absorbs all colors *except* red, which it reflects. You can "see" red because it is being reflected back to you,

The type of light we are in makes a great deal of difference to us. Natural light from the sun is the most beneficial. Artificial lights, in general, function in limited ranges of light wavelengths, and when illuminating an object, allow the object to absorb and reflect only this limited range of color wavelengths. To make a long story short, if you are using artificial lights in place of natural light over long periods of time, you can be depriving yourself of certain wavelengths of color that you need, and may be imbalancing yourself. Some authorities on the subject recommend at least fifteen minutes of natural daylight each day for those who are in artificial light most of the time.

to your eyes.

If you're confused, remember that color is light, white being the presence of all color and black (darkness) being the absence of all light and therefore the absence of color. This is in marked contrast to what you may have learned in Art 101—when "mixing colors," all of the colors mixed together made black or muddy brown. Remember that we are talking about real life here (which is light), not painting (which is pigment or paint).

In general, "things" don't change color, light changes. Let's take a red plastic ball, and place this imaginary ball in an imaginary outdoor setting. As the sun moves across the sky, dips behind a cloud, or is filtered through tree leaves and finally sinks below the horizon, the color of the ball changes. But does it? The ball doesn't *change* color; the wavelengths of light available to be reflected and absorbed become more limited, then less limited as the sun dodges clouds and finally sets. We perceive these different reflections as different colors. Without light, the ball would have no "color" at all. Let's put our red plastic ball in a darkened, sealed room—what color is it? Simply, color is what we perceive between complete darkness and total light. We could not exist in either of these extremes. What is healthful and beneficial lies someplace between the two, in the spectrum of color.

Let's run through the color spectrum. (Perhaps a discussion of colors seems tangential in a book that purports to be about rocks. However, bear with me, and you'll soon see the connection.) I remember a helpful acronym I learned in science class in junior high: Roy G. Biv. Remembering this helpful fellow's name will give you the color spectrum from densest to lightest, from the slowest and most material to fastest and the most ethereal color: *r*ed, *o*range, *y*ellow, *g*reen, *b*lue, *i*ndigo, *v*iolet.

Look at the way the colors are placed on our friend Roy G. Biv. What we have here now is a map of our energy

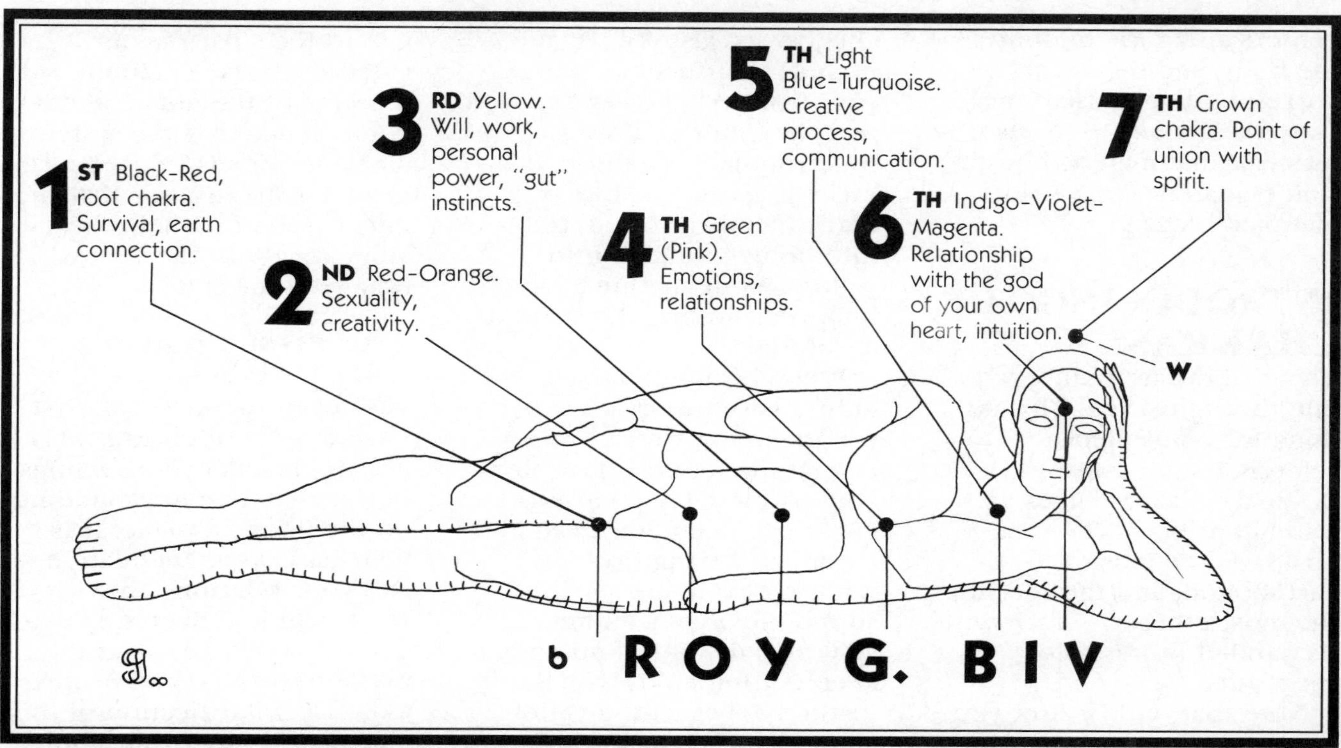

The energy centers (chakras) and their corresponding colors.

centers along the midline of the body, and the color corresponding to that energy center. This *has* to be the ultimate in a matched outfit. Unfortunately, there's no matched luggage.

INTRODUCING THE CHAKRAS

The chakras are numbered one through seven. This is done for convenient referencing, because the original Sanskrit names are difficult at best. The first chakra is the most earthbound, and the seventh the most ethereal—each with its parallel in the color spectrum.

Note that, unlike Roy, you are three-dimensional. Your chakras are also in 3D, and include your back as well as your front. Please note as well that the colors follow a spectrum and are not divided *exactly*, as the chakras are. Black blends into red, red into orange, orange into yellow. As in anything important in life, there are no absolutes.

When healing yourself with colored stones or programming your clear crystal for a color, the color you instinctively choose is the color you need. For example, if you are having problems with work and are suppressing your immune system and you're finding you keep getting colds, you'll instinctively go for a yellow stone, which corresponds not just to the solar plexus (where disease is eliminated) but also to the mind. It is no coincidence that the system-cleaning liver is located near the solar plexus, nor that the citrus fruits that help these ailments are in the yellow-orange color family.

The First Chakra

In the physical body at the base of the spine is the first chakra, or *root* chakra. This also represents your feelings of attachment and grounding on the physical plane. It is your base, your foundation, and your relationship with the world is reflected in this center. (Have a look at the previous ground and protect exercise.) Purification of this chakra will open your ability

Those who are familiar with hatha yoga, or are already familiar with the chakra map, are probably also familiar with the term ***kundalini***, the cosmic life force present in our individual bodies. This is sometimes referred to as "serpent power." Our first chakra is where kundalini resides. We will be talking later about unblocking or opening chakras, even placing energy into them. Please don't confuse this with the actual movement of the kundalini force.

Only after intense discipline and purification is kundalini awakened and brought slowly to the seventh, or crown, chakra. This creates a union of the individual with the all. This is complete spiritual enlightenment and bliss. This state of consciousness has been attained only by the great yogis and saints. Every person has within him- or herself the potential to attain this state, but it is an intense life path, and not suitable for everyone.

We don't have to be perfect to be on the path, however. Anyone who chooses health and is on his or her personal path of healing, or is always striving to be "better," is on this same road. We can look at this path to perfection and purification and venture up the first few steps. We'll be healthier and happier for it. On the way, each step opens us up greater awareness and better control of our lives. A greater ability to create happiness is at our disposal.

You may feel intense joy or other wonderful feelings as you work with crystals and release constrictions from your chakras or as you increase your ability to concentrate. Each of these developmental processes is a step toward awakening

continued on next page

continued from previous page
kundalini, and can be confused with the actual awakening of kundalini. Understand that these joyful, fulfilling feelings and the acquisition of certain psychic gifts are only small benefits along this long path. Great joys and neverending gifts are the fruits on the trees that line the path toward knowing pure spirit.

to have control over your mind. If you do the mind-focusing exercise, you will in effect also be purifying your first chakra. Balancing this chakra will put you at peace with the past, present, and future.

The colors that are related to the root chakra are black, the color of darkness, which is also the color of the feminine principle, and a deep rich red, which is revitalizing and stimulating. Red induces inhalation on a physical level. All shades of gray or smoke-colored stones also work in harmony with this center.

Black symbolizes nothingness and is the origin of light. (Before there is something there is nothing.) Black has been the symbol of sobriety and protection. It has also symbolized from time to time in history death or mourning. It is interesting to note that at different times and in different places around the world, white has also been the accepted color of mourning clothes. Black is not necessarily a "bad" color, but it does have the ability, in great quantities, to put one to sleep or to depress one's mood and physical vitality. It can be a good color for protection because of its association with invisibility.

First-chakra stones are:

- **Clear quartz**
- **Smoky quartz**
- **Black tourmaline**
- **Garnet**
- **Hematite**
- **Pyrite**

The Second Chakra

The lower abdomen, including the sexual organs and kidneys, is the site of the second chakra. This chakra relates to one of the many aspects of our creativity and our vital life force. This is a highly emotional center, and purification and balance in this chakra give you tremendous peace and order in your life. This chakra also relates to your sense of how your higher self relates to your physical, or lower, self. Meditation on this center can purify it greatly and will open your conscious mind to intuitional knowledge and knowledge of astral beings.

Red is the color of this chakra. When working with colored stones, use a deep red one, clearer and brighter than one used for the root chakra. The color might even begin to tend toward orange.

Second-chakra stones are:

- **Clear quartz**
- **Ruby**
- **Chrysocolla—especially for females**
- **Garnet**
- **Carnelian**

The Third Chakra

Located at the navel and slightly above it is the third chakra, which also relates to the solar plexus. Meditation and purification of this chakra eliminates disease from the body. The solar plexus is known as the body's battery, and stores your vital energy. This chakra also has to do with digestion and absorption of food and ideas. The third chakra is the lower "seat" of intuition. We are all familiar with "gut" instincts and feelings. This chakra acts as a bridge between the upper and lower vibrations.

The solar plexus chakra is the color of golden sun rays: orange. This is the color of joy and will elevate a depressed mood. Orange is especially healing to the intestines, pancreas, liver, and kidneys, all located in the area of the third center. Yellow and lighter shades of orange also are third-chakra colors, and tend to relate to the upper half of the physical area, while the orange and darker brownish oranges relate to the lower portion of the chakra. Yellow is the color of the mind, intelligence, clear thinking. Yellow is also a color that works well for healing the large intestine and the stomach. It can also help in treating some headaches.

Third-chakra stones are:

- **Clear quartz**
- **Citrine**
- **Calcite—all colors**
- **Peridot (yellow-green)**
- **Yellow topaz**
- **Amber**

The Fourth Chakra

At the location of the physical heart and the spiritual heart is the fourth chakra. Concentration on this center brings cosmic love and qualities of purity. Psychic powers are gained by meditation upon one's heart.

This is a chakra that deserves a lot of attention. It is where love manifests at all levels. Self-love is located here. So are love of others,

It is OK to place the crystal anywhere on the body you think might answer your needs. Trust yourself. You can hold a crystal in your hands, which tend to be sensitive receivers. Or place it anywhere along the center line of the body (chakra centers), which are also very receptive to crystal vibration. With your knowledge of colors and chakras, you can hold a correspondingly colored stone (or a clear quartz programmed for color) on the appropriate chakra or in the hand.

love of humanity, love of God. The heart chakra is a receiver for vibrations of the frequency of love. As a sender, the heart center sends out vibrations of love. There are no physical limits on this frequency. You can love someone far away, and that person can feel loved.

The color of the heart center is green. Green is the color of balance and renewal. This color harmonizes your body with the color that is found so often in nature and gives a feeling of peace. Green is the color of complete balance. This is an excellent color to use when working with any aspect of balance. Because green's nature is one of balance, the complementary color, red or pink, is often used in healing work. This uses the stimulating action of the reds. Pink also gives a feeling of contentment and helps one feel the joy of activity. The watermelon tourmaline is an excellent heart-center stone. It is inherently balanced with the red and green vibrations, as the name suggests.

Fourth-chakra stones are:

- **Clear quartz**
- **Rose quartz**
- **Green tourmaline**
- **Watermelon tourmaline**
- **Pink tourmaline**
- **Malachite**
- **Emerald**
- **Pearl**
- **Kunzite**

The Fifth Chakra

Fifth is the throat chakra. This is the other center of creativity. It is reaching out to others. This is sometimes called the communications center, and it is not coincidental that the voice comes from this center. Meditation on this center brings great success in this area on the material plane and on the higher vibrational planes. This center of communication is often regarded as the artist's chakra.

The colors to work with when you are concentrating on this chakra are turquoise and light to medium blue. Blue in the lighter shades is cool, refreshing, and calming. This vibration is particularly important for people who endure a lot of pressure in their lives. Blue gives a feeling of unwinding and induces a controlled exhalation. In our busy, stress-filled world, all ranges of blue are healthful colors, especially where tensions and limitations cause feelings of frustration, unease and ill health.

The deeper blue colors, especially when working with colored stones, tend to be useful with the sixth chakra rather than the fifth. This begins to define a rather fine line and individual cases must be decided instinctively; there is no general rule.

Fifth-chakra stones are:

- **Clear quartz**
- **Aquamarine**
- **Turquoise**
- **Chrysocolla**
- **Blue topaz**
- **Blue lace agate**

The Sixth Chakra

Another name for the sixth chakra is the third eye. This is located between the eyebrows and is represented on the physical level by the brain. The sixth chakra is the seat of the mind. Concentration on this chakra will bring intuitional gifts and have a positive effect on past-life karma.

The sixth center is associated with the indigo (dark blue) to violet range of the spectrum, which includes magenta. This range of colors may seem too broad to be represented within one center, but if you look at the color spectrum as projected by a prism, the blue-to-magenta range is rather narrow. The vibrations of these colors are very fast. As the vibrations increase, the change in color is noticeable, blue to violet to magenta.

Each color change is a slight variation on a theme, and corresponds to individual reactions to spirituality. The deep blue ranges of color can be used to give deep peace and rest to a person's body and soul. Violet is the balance of red and blue together. This brings inner balance—

calming to the body and heightening to the senses. Violet brings peace combined with activity; it is the color of prayer and meditation. It offers balance that is similar to green. Violet brings you into balance with creative forces outside yourself. Violet is helpful in healing therapies that are treating the heart, lungs, and blood, especially where there is some sort of infection.

Sixth-chakra stones are:

- **Clear quartz**
- **Amethyst**
- **Sapphire**
- **Lapis**
- **Moonstone**

The Seventh Chakra

Also called the crown, the seventh chakra is not located on the physical body but does correspond to the pineal gland. The pineal gland is a small gland in the brain and is part of every vertebrate having a skull. Its function is unknown; it is supposed to be a gland that was once a sensory organ but is no longer functioning. The seventh chakra is associated with deep religious understanding and oneness with the universe.

The color associated with the seventh chakra is white, or the color that includes all color. White has long symbolized purity. It is the preferred color of protection. Of course, all other centers are intimately connected with the crown chakra.

The stones that correspond with this center of mystical origin are clear stones, the stones that have all color vibrations included in their spectrum. When working with this chakra, the quartz is the stone of first choice.

Seventh-chakra stones are:

- **Clear quartz**
- **Zircon**
- **Diamond**
- **Herkimer**

8 ◆ BALANCE AND HEALING

It was a wise person who advocated moderation in all things. The same goes for our chakras and every aspect of ourselves; balance and harmony equal health and happiness.

One of the great attributes of quartz is that it can help you balance your inner being, and unify your emotional and your mental being with your physical being. Imagine yourself in three transparent layers, and imagine your seven chakras. Now imagine these layers not overlapping properly, and imagine what this does to your chakras. Maybe five are glowing with energy while two are much dimmer. Perhaps there is a splotch of constriction here or there. Maybe one chakra is glowing like mad, a regular searchlight.

Let's take this image and pack the body images, emotional, mental, and physical, so they overlap in sync. Let's turn the lights of the chakras up or down so they are in an even harmony. This is balance. This is the way you function at your optimum. Each part of you takes its correct function. It would seem silly to allow your arms to function as your legs; they weren't meant to take you from place to place. It is just as absurd to allow your heart to reconcile your checkbook or to let your mind choose a friend. Correcting any of these imbalances is one aspect of harmony. Uncorrected, they can lead to physical, psychological, or emotional disease or dis-ease.

The other relationship that must be harmonious for your complete balance is the relationship between your insides and the outside. It is important that you do not feel isolated, but that you are a part of something larger than yourself. You can often find this through a relationship with another person or an organization of

As a crystal user, you can move into close communication with "cosmic" energy and so to the god of your own heart, which results in a refining and increasing of extra sensory perception (ESP). This usually manifests as an increase of your intuition, allowing you to make better "guesses" as to what roads will be freer of traffic, what time to get home in order to receive that important phone call, or how much food to buy at the market because you "sense" that two dear friends will drop by near dinnertime. These types of coincidences may happen more frequently as you become more balanced and more open. It's an indication that your perceptive abilities are increasing.

When you go ahead and flow with your feelings, when you take that road your heart or gut is telling you about, when you buy the extra food, these are indications of your progress. This progress is your trust in you. It shows that you are trusting yourself and trusting the universe.

some sort, but the relationship I'm trying to get at goes deeper, to your core. When the party is all over and everyone goes home and you close your eyes and are seeing the backside of your very own eyelids, this is when you should feel the harmony of being integrated with something bigger than yourself, something outside yourself. Whether you relate to the outside through your head or heart or root chakra is highly individual, but everyone does have a unique place of balance.

The universe will always seek balance. To counteract the flow of universal energy is not to be feared necessarily, but it is futile. To try to keep energies off balance for extended periods may be successful in the short run, but in the long run, correction is inevitable. Balance is effortless; it is not a point, but a flow. Balance changes because life changes. And balance is nonjudgmental.

Now this is not to imply that you should just flow along and aimlessly bump from shore to shore, so to speak. Of course, you should take control of your own life, make decisions and choices. These are personal and individual and should be satisfying to you. Find your own balance.

You actually have the ability to find your own balance, to heal yourself of any ailment, to change. If you are sufficiently motivated and dedicated to change, you *will* change. Your ability to change and your ability to heal yourself are a direct result of your motivation to accomplish. *With the proper effort, you* will *succeed.* If you want to change the shape of your nose, with sufficient effort, you can. (The amount of psychic effort and time, of course, would be enormous. But if this is your goal in life, go ahead.)

MAKING THE RIGHT CHOICES

Unfortunately, many of us make choices that throw our physical and energy bodies out of alignment. We work too hard and get too little sleep; we eat too much and get fat; we pick fights with our lovers and throw our love centers out of balance. This will often cause one or more chakras to be out of whack. Usually, of course, the natural balance of things will take over. You get sick and are forced to stay home from work and get some rest; you start feeling fat enough to diet and get in better shape with exercise; you realize that you must resolve your differences with your lover. We can resist these healthy, balancing instincts, but this resistance to change, stagnation, refusal to flow with the natural path of energy, can result in deep unhappiness and even physical disease. Unhappiness and disease are the body's way of signaling our resistance to change or neglect of ourselves. What we usually perceive as "being sick" is really our body taking steps to get us well. Unhappiness is your emotions saying "something is going wrong." Causing an argument is an attempt to resolve an issue.

The most important healer for you is you. It is the natural tendency of your body (this is you) to seek health at all levels.

If you are uncertain and confused and can't decide about one thing or another, your crystal will amplify these scattered thoughts. If this is your state of mind, you can meditate with your crystal and ask yourself what is the best answer to a particular set of circumstances, or how best to center yourself. Otherwise, put the crystal away until you are really ready to focus.

Here's an exercise to do with your crystal that can help you make these decisions for yourself, when you have a serious question in mind—a career change or a love-life question. Understand, this is not a parlor game, but a way of accessing information that is deep inside yourself. It is always there, but may be clouded with the confusion of the moment. This exercise can help you understand consciously what you really want in any given situation.

Exercise 7: Decision Making

1. Find a quiet place where you can sit undisturbed for fifteen minutes or so. If you have a regular place of meditation or deep concentration, go there. Try not to be under a time limit.
2. Do a quickie ground-and-protect or, if you have time, go deeper into the exercise. This increases receptivity and also keeps you on balance (or helps you find balance, whatever the case may be).
3. Now, place your crystal in your receiving hand (left for righties) and hold it vertically so that the point is up. If you are not particularly drawn to hold it your own special way, you might try holding it along the midline of your body near the heart center. Another way of holding the crystal, which might feel more natural, is in the palm of your hand, point away from you, and perhaps against one of your fingertips. Relax your arm, resting on your knee, for example.
4. Tell yourself now that you know you will find the answer to your question within yourself. This answer will be on the path of balance and flow from within.
5. Ask your question now; as you do so, you can gently rub the crystal if you wish, although this is not absolutely necessary. As you ask the question inwardly, focus your mind on one of your chakra points. Choose any; the heart center or the sixth (brow) chakra are both very useful in

this type of meditation. But once you choose one for a particular meditation, don't flip aimlessly back and forth. Concentrate on your question and the chakra.

6. Listen.
7. When you feel satisfied, place the crystal down on a nearby surface and take a moment to breathe and reenter your usual day. You may want to say a short closing prayer, with any sentiment you find appropriate.

Part of selecting your priorities is selecting what messages you send yourself. Your thoughts do come back to you. When you snarl and complain and figure everything is against you, it's no mystery when you step off the curb and hit something slippery and break your leg. "Well, isn't it just my luck?" you say with conviction. When you project love to yourself and others, love comes back to you, doubled, tripled, and squared.

Stilling the mind as in Exercise 5 is a giant step in taking control of your thoughts. Stilling your mind increases your ability to concentrate when you are engaged in some thought activity, because you are not being constantly bombarded with unwanted thoughts. This is very important in work with crystals, because, as I mentioned, a crystal will amplify and project your thoughts. If the thoughts are scattered, unfocused, and unclear, this is what you will project, and this is what you will be getting back.

Often, what is standing between you and healthier or happier existence is some thought pattern that is unproductive or is more seriously destructive. If you believe you are a person stuck in a career with limited growth potential and there is no way out, then you are. If you believe yourself to have tremendous potential, then you are sure to find an opportunity and make the best of it. The universe is a place of abundance and strength. If there is something you are looking for or need and can't find it, maybe you are then looking in the wrong place. Negative thought patterns can be found lurking around where there are blocks in your life, or where you feel things could be working better. Be watchful of your thoughts. Sit with your crystal, and listen and find out where these patterns exist. Take these patterns and turn them around, or just kick them out with the next exercise.

Please note that balance is not a static state. To stay in balance, you should not strive to remain impervious to forces around you; rather, you should be at peace with them. For example, a boat may be balancing itself nicely in the water, but if a storm comes up, it must "roll" with the waves to stay afloat. You, like that boat, must stay in tune with your surroundings to stay balanced.

Exercise 8: Affirmations for Balance

1. The first step is to catch the repetitive chatter in the back of your mind—those nagging thoughts. Jot them down on a piece of paper. Be especially alert for phrases that begin "I never . . . ," "Whatever made me think . . . ," or "I'm always . . . (negative, negative)."
2. If you can't find the block to a particular situation you're unhappy with, do the exercise related to decision making. Ask your block-related question, and listen to your own heart. You are the most important healer of yourself.
3. Once you've found some material to work with, begin your turnaround. Take your negative thoughts and make them into positive thoughts. When you have a phrase that starts "I never . . . ," simply turn it into "I always." For "Whatever made me think . . . ," make it, "I always make the right decision," and so on. Even if you can't quite believe it at first, say it anyway. You'll get used to being right; you'll get used to praising yourself to yourself. A wonderful phrase to say to yourself might be "Good things always happen to me." You might think this is awfully Pollyannaish. My only response to that is that you'll get over it. Give it a try! And by all means have fun with turning your negatives into positives, and kiss the negatives good-bye, because you won't be seeing them for long.
4. Once you have your list of positive thoughts, you can repeat one or two to yourself while you sit in deep concentration with your crystal (again, as in the decision-making exercise). Or put your affirmative thoughts on a personal tape—get your portable player with your headphones, take your crystal in your hand, and head off for a walk in the park.

PROGRAMMING YOUR CRYSTAL

Just as you can use your crystal actively or passively, you can also use it "programmed" or at its natural frequency. At its own frequency, it's best for

- *Helping to clarify thought*
- *Making decisions*
- *Figuring out what areas of your life need work*
- *Deciding how quartz can best heal you; deciding what to program for*

A programmed crystal, on the other hand, is what you use when you are actually working on a specific problem or goal. Once you program a crystal to oscillate or vibrate at a certain frequency, it will keep on vibrating at that frequency. Once you've programmed it (as you did earlier with the "automatic clean" program), it will continue, without further input from you, to do the good for which it was programmed. For example, if your Uncle Joe is in the hospital recovering from an appendectomy, you could program your crystal with the visualization of Uncle Joe sitting in his garden in the sunshine, as happy as could be, enjoying perfect health. Now you are a busy person, so you spend ten to fifteen minutes on the program, set the crystal on the mantlepiece, and go off to work. Of course, you won't forget to call Uncle Joe, but all day long the crystal sends him your healing thoughts. You may even decide to drop a programmed crystal off at the hospital for him to have. Or do both.

The possibilities for programming are endless. You could program:

- *A thought of love or health for yourself or another person or the world—the thought will continue to radiate like a beacon.*
- *A prayer*
- *A feeling of confidence, stored for a moment when you might be feeling low*
- *Any positive thought that might relate to any area of your life that you may be working on at the moment*

- *To open any blocked area—you could program to open your heart chakra, for instance, if you feel you've not been as loving as you've wanted to be lately.*
- *For a particular color—if you feel in need of a lift, some stimulation, program red. If you are feeling blue, program pink. Feeling nervous or frenzied? Play it cool with blue. The quartz can be programmed with any individual color because it contains all colors within its full spectrum of white light. You may want to program a color to correspond to a particular chakra.*

The following exercises show how to program your crystal. Before programming, you might want to make sure you are starting with a "clear" crystal. As always, use your intuition. There is no limit to what you can program.

You can use your programmed quartzes the same way you use a "natural" stone. The colored programs are good in the chakra-opening exercise that will follow, or it is fine if you want to use this stone in a pattern as Joan did earlier on in the book. You might want to program a quartz as your bedside companion to give you restful sleep or to help you wake up refreshed. Don't be upset if you cannot readily

Exercise 9: Programming a Crystal

To introduce the program you wish, place the crystal in your giving hand (right for righties) and, with *very clear intention*, send your thoughts through your arm along your energy body into your hand, and imprint this in the crystal. That is all. Anytime you wish to receive these thoughts, put the crystal in your receiving hand, and just be receptive to it.

Exercise 10: Programming a Crystal with Color

There are two ways to program your crystal with color:

1. The first is as you would program the crystal for anything else. As you gather your intent of thought, the thought should be of the color. Visualize the color, look at the color if possible, and think of all the positive attributes of the color as you are programming.
2. The other way to program a color into your crystal is with the surface on which it rests. After clearing the crystal, as you are charging it in the sun or moon, instead of placing the quartz on a white surface, choose a colored surface. Use the color you want to infuse into your quartz.

pick up the programmed thought vibrations from your crystal, at least on a conscious level. If you want to develop this particular psychic ability, you can with some practice.

One more suggestion: If you're going into a situation that is familiar, but you are anticipating particular undesirable interactions, program your crystal for change. Program your crystal to make you and events in your life what you want them to be. You can project into the future and make things right before they go wrong. Bring this crystal with you and relax, knowing that the change you want to effect is for positive growth. It will take place with ease. There is no limit to the positive change you can put into effect.

HEALING WITH YOUR CRYSTAL

To me, *healing* is synonymous with *balancing*; if you are balanced, you will be healed, and if you heal yourself, you will be balanced. However, many people are used to thinking of particular physical ailments as discrete, and they find it difficult to realize that balancing will actually iron out many physical problems. It is important to work on understanding the connection between healing and balance, as it will enable you to better use your crystal to effect positive change. It will also help you feel connected to the wholeness of the world, as we discussed at the beginning of the chapter.

Generally, you will want to "heal" yourself with a crystal cleansed and programmed for the particular type of goal—overall cleaning, particular chakra unblocking, whatever. I sometimes suggest to people that they keep several crystals around—one for cleaning, one for unblocking, one kept at its natural frequency for decision making, and so on.

Have fun with your crystal! Don't be afraid to jump in and start. If some of this seems confusing to you, don't worry about it now. The more you use your crystal, the more you will understand. Just think of crystal knowledge as a circle—there's no beginning and no end. Begin wherever you feel most comfortable, and remember:

- **Don't postpone joy. It's yours now if you want it.**
- **The path of health and healing is full of acceptance. As soon as you feel you are denying yourself anything for the sake of improvement, it's time to inspect your modus operandi.**
- **You can make things right before they go wrong by anticipating the best.**
- **Listen for the message in your own heart. You will know when you have found it, because it is the voice that always brings happiness.**

Exercise 11: Cleaning Your Energy Field

You've come home from a long day at work or you've just finished the run to the vet for your cat Spot, and after a long line at the grocery store, you are *ready* for a cleaning. You've gone through six months of crown work at the dentist, and you are ready. The landlord "forgot" to fix the sink again, and you *need* to clean your energy field.

1. Have your crystal handy.
2. Ground yourself (see Exercise 6 if you need a review). Sit and take a few breaths, and from the bottom of your spine, send a grounding line to the earth. Feel your earth connection. From the top of your head, send a line up to the sky and hook up to the sun, the source of light. Take a few more breaths. Use each moment; be aware of what you are doing at every moment. Focus your intention on the moment.
3. With open eyes, take your crystal in your hand. Holding it so the broad side faces out, go slowly along your energy field, starting with your shoulders, and "brush" your arms, down to your fingertips. Then start at the other shoulder. You may want to shake the crystal off from time to time. To do this, just pretend you are shaking down a thermometer and shake the constricting energy off the crystal. You can also use this flicking motion in cleaning off the energy field with the crystal. Try them both.
4. After clearing your arms, use the broad side of the crystal and go over the surface of your head and face, about three to five inches away from your physical body. Have your intention in mind: You are cleansing your energy body of the day's events. Use the crystal as if it were a bar of

soap, lightly covering the entire surface. You don't need to scrub—passing over your energy body should be sufficient. In an area that you feel needs more attention, try leaving the crystal a little longer. Or gently pulse the crystal back and forth, as if you were beating a small drum. You could also place the nonterminated end of the crystal toward the body at the place that you feel is the center of the problem. A little bit of focused attention goes a long way; five minutes is a lot of time for any one area. Keep the crystal moving.

5. Now go to the torso and work in zigzags. Cover the whole body area; once you've worked down the body, try not to work your way up again. Once the process of cleaning has begun, try to see it through in a rhythmic manner.
6. Once you've done the entire trunk of your body, following your intuition at every moment, work your way down one thigh, then the other. Then do each calf and foot.
7. Shake off the crystal. Put it down, and shake off your hands. You may want to yawn or stretch or shake yourself off a minute.
8. Take a seat, and renew your grounded feeling. Take a breath or two, and feel the lightness and the calm. Imagine now that with every breath you are bringing energy-giving light into your body and are exhaling worry, sorrow, grief, pain—with every exhalation, you lighten your spirit. Inhale and invite order; exhale that gray fog of stress. Exhale, releasing guilt and judgment; inhale, bringing in love and acceptance.
9. Sit for a few moments. At this point, you could choose to go further into prayer or meditation or set your mind to gear into the physical world of tasks. Either way, you will feel refreshed and strong after this self-healing treatment.

Exercise 12: Cleaning and Opening Your Energy Centers

1. Sit comfortably and have your crystal nearby. Focus on your breath; do a quickie ground-and-protect.
2. Gently hold your crystal in your hands with the termination point away from you. Bring the back end of the crystal to the root chakra at about three to five inches from the physical body. Bring your attention to that area. Stay there for a few moments—at least forty-five seconds but not more than five minutes. Let your mind flow; if you find your mind at the doughnut shop, bring it gently to the matter at hand. Let associations in your mind develop. Information may come up for you to sort through. This will open up the energy centers, and you might go through a good deal of inspection of past events that you have stored away in the form of little constrictions. Allow them to open slowly and at a pace you feel comfortable with.
3. Repeat with each chakra. Some chakra areas are more sensitive than others. Be sensitive and kind to yourself. If you feel your heart area is very sensitive and you don't feel comfortable giving it as much intention time as other areas, *that's OK*. Anything is OK. Follow your inner voice. Trust yourself. Follow along the entire body, stopping at every chakra: root, reproductive area, solar plexus, heart, throat, brow, crown.
4. After you have opened every chakra, sit and enjoy the serenity of the moment. Enjoy the feeling of being as clean and open as you are at this moment. Imagine white light coming into each center. Perhaps you feel one center in particular—use that one center to draw in light, and

allow it to flow all over your system, bringing expanding and healing white light to any areas of your body where there is constriction, pain, or disease. Let the white light penetrate this malady and break it up. Imagine light scrubbing out the darkness of pain, dissolving it and letting it flow, dispersing it into your system and gently moving it away—a soothing balm.

5. Breathe a few more breaths, and with these breaths bring yourself to focus on the room you are in and the time of day. Prepare yourself to resume the rest of your day or evening. If you feel you might be too "open" and want some protection, you can cloak your energy field with a black shell or sunscreen, as you did in the grounding exercise. Just create it in your mind, and it is there. Or you may, under certain circumstances, close down your energy centers slightly. Use your own judgment, taking into consideration what you will be doing and how you are feeling. If you are at the end of the day and all you'll be doing is relaxing, you probably want to say a prayer and continue relaxing. If you leave the session to encounter an urban situation, you may want to put up a protective colored covering on your energy field. Some examples might be dark blue (for calming), light blue (for joyous feelings), yellow (alertness), orange (also joyful, but in a more active way), green (balancing), and violet (to inspire a meditative outlook).

***Note:* When a negative has clung so tightly and deeply that it has manifested itself as a physical ailment, it can take longer to dissipate, and very often requires physical attention. It is *important* that you seek the *appropriate professional treatment* in these cases, as well as continue to work with the energy field.**

One of the unintentional benefits of opening and aligning your chakras may be that you'll start to develop psychic tendencies. You may find you are receiving flashes of insight or even vivid images from the past, present, or future that you can't explain. Don't worry. This is normal. The more you work with your chakras, the more you realize that anything is possible. When you release the negatives that stop you from believing, you are unblocking routes in your path to understanding. Psychic experiences are part of this understanding.

I have one last exercise for you that is simple but terrifically effective. It involves visualization, or seeing an image of a goal you have for yourself or the world realized in your mind's eye. As we've seen, thoughts or intentions are very powerful. With goal-directed thoughts or intentions running through your mind and being amplified by your quartz, you can speed up your healing process. Visualization and your crystal are greater than the sum of their parts.

You can perform this exercise on yourself or on another person. When you work on another, *always* remember to ground and protect yourself. Also always ask permission; be sure your subject *wants* to be healed.

Exercise 13: Healing Specific Ailments

Imagine a particular area of your body requires greater health. Don't forget to ground and protect yourself!

1. Hold the point of your programmed crystal toward the ailment.
2. Envision white light (or any healing color ray) running through the crystal and permeating the area of pain or dis-ease.
3. Imagine the area in perfect health, and hold this image for a half-minute or more. During this time, you can imagine waves of healing energy going through your arm, through the crystal, motivated by your conscious breath.
4. Put down the crystal. Breathe fully and gently. Continue to imagine your breath flowing to the area you are working on.
5. Allow the light you have envisioned flowing to the area to slowly fade. The pain or dis-ease fades with the light.

9 ♦ A GUIDE TO BUYING HEALING STONES

Buying healing stones is a bit different from selecting stones for any other purpose. You can purchase stones for investment, which supposes that you will eventually resell the gem for profit. Or you can buy a stone as a geological specimen, which generally means that you are collecting your specimens with some idea or category in mind. You might have rarity in mind or some other special quality.

Healing stones are often very ordinary stones. One can have a collection of excellent healing stones yet, in the eyes of a collector or investor, not have much of anything.

None of the usual sources or guidelines are necessarily valid when it comes to stones for healing. To choose the proper stones for yourself and your family, simply trust yourself. Even if you don't know *anything* about stones or healing, *trust* yourself. Do some reading and some looking. Trust yourself to know what is best for you. You know best what is compatible with you.

I've had a lot of experience helping novices select stones for themselves. Ten times out of eleven, the person has already selected the type of stone and is only seeking validation of that intuitive selection.

You know more than you think you do. Every day, at least one person comes up to me saying, "I don't know what stone to use for my stomach," yet that person is invariably holding a citrine, or hovering over a shelf full of them. Citrine, of course, is just the stone he or she needs. Other crystal healers tell similar stories. If a person is drawn to a stone, he or she needs that stone. If you feel that you should place your crystal over your abdomen, that is where you need it. Just give it a try. You can't do any harm, and you could do a lot of good.

POINTERS FOR SHOPPING

Keep the following points in mind when you are looking at healing stones for purchase:

- *You will automatically be attracted to the stone group and the particular stone that will be of value to you in your work with healing stones.*
- *All price ranges are available within the different mineral groupings. Even an emerald can be had at an affordable price. You don't have to spend a fortune!*
- *A gem can be tiny and do its job. Size is not equivalent to power in the world of minerals. Some small stones can be as effective as larger stones. Just because a stone is large does not necessarily mean that it has more healing power. Go to the stone you are drawn to, but don't forget to keep your particular budget in mind.*
- *Choose a stone for its vibration, and don't be concerned whether the*

The point of a crystal sends energy; the butt end of a single-terminated crystal receives or draws energy. However, energy radiates from the whole crystal—top, bottom, sides. It is in the direct channeling of energy that the point is important. A whole point is sometimes better, but, just as often, the crystal's own energy body "remembers" that the point was there and works just as effectively if it is chipped or broken.

stone is beautiful in the usual sense of the word. The stone you are drawn to may or may not be a desirable stone to the collector or investor, but it may be very suitable to the healer. If you choose the correct vibration for you, chances are it will be pleasing to your *eye. This does not mean that a classically beautiful stone is empty of healing properties. Some stones, however, will feel "empty." This holds true of very polished stones as well as the rough crystals and stones. There are no hard and set rules. Intuition is your surest guide.*

- Please *don't buy a stone that you have negative feelings about. It is a waste of time and money.*
- *It's important to understand that a crystal is a unique item. There is only one exactly like it on earth. If you find what you want or need, my advice is to grab it! Crystals are not widgets that you can order in a particular size and shape from the widget factory. There is no guarantee that a shopkeeper can find*

"another one" of any particular stone. This may seem self-evident and simple, but my experience with healing stones shows me that this concept is difficult to grasp.

- *You might want to wear healing stones as jewelry. Crystal jewelry is available in some stores, and the traditional stone jewelry you see in other stones also possesses healing attributes.*

TYPES OF STONES

A number of different types of stones are available. Some stones are cut and polished, and others are rough. Let's look at the different stone forms available for purchase.

- *Uncut, Unpolished Crystal Form*

Your quartz crystal falls into the category of uncut, unpolished crystal. It has come from the mine; it may have had a bath in oxalic acid to remove the dirt and some frosting or iron that may be present on the surface of the crystal. Otherwise, it has not been changed.

- *Polished Crystal*

In contrast, a polished crystal has basically retained its natural crystal shape but has been polished along each existing crystal plane. Generally the butt end of the crystal (the broken end where it had formerly been attached to its matrix or mother rock) has been flattened or rounded. These changes will remove any imperfections found on the surface of a crystal. This can often improve the healing properties of a stone, if done with great care and consciousness.

- *Manufactured Crystal Shapes*

Gem cutters often form stones into crystallike shapes that are not their natural shape. For example, a piece of rose quartz may be fashioned into the shape of a polished natural crystal, perhaps the shape of a clear quartz.

• *Spheres*

Stone cutters shape spheres of stones, but these are not natural shapes. These spheres have been used for scrying (see the earlier discussion of history), and this is a separate technique from the uses of stones for healing. The sphere as a healing stone emanates a slightly different vibration than the stone in its natural state. The vibration is dispersed in direct relation to the shape. The sphere will emanate along the entire surface and pulse in a spherical shape. This can often soften the energy of the stone, which can be of great value.

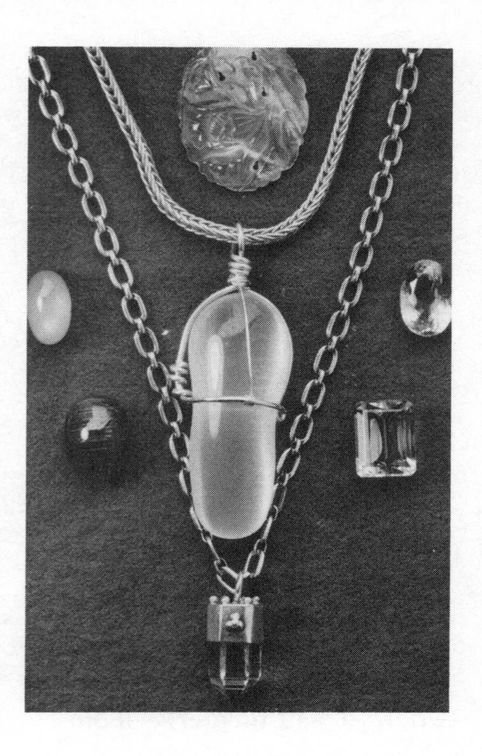

• *Egg Shapes*

Egg shapes follow the general rule for spheres. The egg shape is popular (especially with chickens!) and does have the psychological message of a promise of new birth.

Healing stones can also be worn as jewelry.

- ***Obelisks and Other Monument Shapes***

Stones shaped like obelisks and other monuments are wonderful display items. I believe these are especially suited to passive use.

- ***Rough Stones***

As they come from the mine, many stones do not have large crystal structures. Rose quartz, bloodstone, carnelian, and many others are formed in massive veins. When miners take them from the earth, they break off chunks of the stones. You can buy "rough" in many shapes and sizes.

- ***Cut, Polished, Faceted Stones***

Cut and polished stones and faceted stones are the traditional stone shapes you are used to seeing in jewelry. The terms are somewhat self-explanatory.

10 ♦ A BRIEF GUIDE TO COLORED STONES

The following pages give useful and interesting information about many stones that are being used and recommended by today's healers.

In using the colored stones, you might find that the first place you may feel the stone is in your heart area. This is because the heart center acts a bridge between the physical being and the spirit.

The information presented here is meant as a starting point. The relationship you have with each different crystal type is unique because you are as unique as each crystal. There will also be variations from crystal to crystal. Go ahead and experiment—above all, *have fun*.

TYPES OF COLORED STONES

Amethyst

The amethyst is a lilac to deep purple quartz stone. This type of quartz, with its beautiful terminated crystals, has long been used as an amulet to prevent drunkenness. The reason is that this stone is a powerful sixth-chakra stone. The attributes of this chakra are the ability of getting in close touch with the cosmic powers and the god of one's own heart.

Amethyst is stimulating to the heart and to the blood and purifies both. This stone will bring about changes toward the purity of heart and mind. This is a wonderful meditation stone, especially if your meditations are asking questions that have to do with spiritual matters or questions that need attention by your spiritual guides. Amethyst helps to develop psychic vision and is useful in treating color blindness.

Aquamarine

Often referred to as "aqua" for short, aquamarine was given this name because of its color and its resemblance to the sea. It was known long ago as the stone of seamen, and it remains a protective stone for all travelers, especially those traveling across water by boat or air. Aqua was also frequently given as a token of friendship and fidelity. Its uses today expand on these meanings and attributes. Aqua helps bring about a feeling of calm and serenity. It is also a stone of communication and encourages joy and peace, with an accent on joy. This is truly a chocolate-chip cookie on a rainy day. This stone is similar in vibration to that of turquoise; both stones help you speak from the heart.

Azurite/Malachite

The commonly found mix of azurite and malachite is greater than the sum of its parts. It is excellent in the treatment of maladies that have gone deep into one's body and psyche.

Bloodstone

The "free ride stone" is the nickname for bloodstone, because it just makes the pain go away, without your having to confront it again. You only have to be near this stone. It will take away long-standing sadness and soothe memories that make the heart sore.

Carnelian

Used throughout the ages by the Egyptians and the Persians, the beautiful earthy red carnelian is a grounding stone and can deepen one's appreciation for the earth and all its beauty.

Citrine

The stone of the solar plexus, citrine is the storehouse of your personal power and a center of your intelligence. This stone is a color of joy and brings light and sunshine into your life in the form of easily used energy. Stimulating to the intellect, this stone is good for sharpness of wit. It is also of use when approaching any issue that has to do with digestion or assimilation, or integrating change into oneself. Citrine also stirs up creativity and aids in decisions; it is particularly good for business decisions.

Diamond

The diamond is a symbol of duration in our society, because it is the hardest and therefore the most durable of all gems. This stone relates to the seventh chakra, the crown. Diamond is an amplifier and a purifier.

Emerald

Another heart stone, the emerald represents the pinnacle of the green healing ray. It acts as a balancer and purifier. It is also *the* stone of abundance.

The Garnets

The *green garnet* is a stone of the eyes. It will help dissolve any visual problems and any irritations. It will also dissolve bitterness anywhere it manifests in the body, working through the eyes.

The beautiful *red garnet* is a balancer and purifier to the second chakra. It will place the sexual emotions into their correct perspective, harmonizing the entire system. This stone also will inspire deep love and devotion.

Rose quartz is always beneficial. The frequency of love given off by the color pink is always helpful when there is discomfort present. Rose quartz can almost always be used in conjunction with other stones. The vibration of rose quartz can have a modifying or softening effect on other stones, as well, which is especially useful for "blending" if you are using a combination of stones.

Hematite

The hematite is an important stone for those who are space cadets first class. It will help people who are scattered or often light-headed to keep themselves together. It will also help, when you tend to be very affected by crystals, to bring you back to a more centered awareness. Used as a meditative stone, hematite grounds information through the crown chakra and energies from other realities into this physical reality. Hematite is good for any blood disorders, especially where the blood has little clotting ability.

Indicolite

Blue in color, the indicolite tourmaline relates to the throat and the brow centers. It seems to carry a strong green ray as well, because it affects the heart center, too. This stone seems to connect all three centers, balances their functions, and integrates one's intuition with the heart and makes this available for communication.

Lapis

A deep blue stone, lapis is connected to the sixth chakra. A distinguishing characteristic of lapis is its gold specks, which are not actually gold, but the "fool's gold"—pyrite. This stone was used extensively by the Egyptians. Like turquoise, it can be used as a ancestor connection for those so inclined.

Malachite

An important stone on a physical level, malachite can be taped to sprains, aches, bruises, or cuts. Healing will occur faster, and this stone removes pain in an absorptive manner. It is important to clean this stone thoroughly after treatments. Malachite is also known as the "male warrior stone" and can bring assertiveness and a greater sense of self to the user.

Moonstone

The moonstone is a stone of the moon energy and amplifies and helps reflect the emotional state. It is a stone that increases the intuition. It can be a good stone for those who have difficulty getting in touch with their emotions. Some women might want to set this stone aside at certain times of their cycle.

Obsidian

Another important stone for our times, obsidian helps you look at things that are blocking you from your spiritual growth and development. This stone requires a fair amount of cleaning. Since it acts as a mirror for what is deep inside yourself, obsidian should be used at appropriate times and set aside when you are through. Give yourself a dose of rose quartz and/or green tourmaline after working with obsidian. On a physical level, obsidian can have a drawing effect when applied to a pain or sore.

Pearl

Not a gemstone from the earth, the pearl is created by oysters and other mollusks, creatures of salt water and fresh water. The pearl can soothe the emotions of the user and gently purify the emotions through the heart.

Pyrite

Another important grounder, pyrite can also be used as a "money magnet." Once known as fool's gold, pyrite can help attract the real thing.

Rose Quartz

Part of the quartz family, rose quartz sends and receives on the universal frequency of unconditional love. This is sometimes known as the friendship stone. It is always a gift of love, for sending or receiving. This is a good stone for meditation on love of self, love of others, or sending love to planet Earth. Health can be sent to others on this frequency, and love in any form to anyone. Primarily a heart stone, rose quartz heals one's own heart and helps you see opportunities available for positive growth. It is a great stone to include if you are using more than one stone to heal; it helps stones work together harmoniously.

Rubilite

Highly prized for its color, rubilite is a pink to deep pink tourmaline. This is a stone of the fifth chakra. It is a valuable heart opener and cleanser of old sorrows. It is a color that approaches the violets and therefore also helps bring intuitional knowledge into the heart. It also opens one to intuitional love.

Ruby

The ruby is a heart center stone that acts on the emotions and the physical heart. It strengthens and stimulates and will act on the entire system because it purifies and vitalizes the blood. Ruby is thought to strengthen the will and thereby give courage to the wearer. Ruby will also act on the second chakra, bringing balance and clarity to issues of polarity (male and female).

Rutilated Quartz

Clear quartz or smoky quartz with inclusions of rutile is called rutilated quartz. This stone functions much the same as quartz without the inclusions. Rutile makes the stone excellent in getting rid of all types of skin infections, such as rashes and acne, and also helpful after sustaining a cut or burn. This is because it can go deeper than the skin and removes deep infections. It also breaks up the block that gets in the way of creativity.

Sapphire

Considered a "precious" stone because it is rare, sapphire is a stone of wisdom and intuition. It relates to the sixth chakra. While known for their indigo blue, sapphires are mined in every color of the rainbow. This stone seems to relate to the brow center, regardless of the color. The sapphire brings a feeling of lightness and joy to the wearer. This stone can bring clarity and depth to one's thoughts. In the deeper blues to the opaque colors, this gem also affords the wearer protection from negativity.

Smoky Quartz

The smoky quartz has a grounding nature. Often used for the root chakra, this stone can be worn anywhere on the body and has an extra soothing and calming effect good for those of a strongly active personality. It is helpful in times when fearful feelings enter one's mind and spirit. Like rose quartz, this smoky gem can be used when working with a number of different stones to balance the energies of the stones.

Sodalite

An important New Age stone, sodalite resembles lapis, although it has no pyrite and is sometimes veined with white. Sodalite is an absorptive stone and is important in the treatment of all diseases and unwanted habits. It will detoxify the system and is important in the relief of fear of disease. To work this stone, hold it in your hand and rub it. Or carry the stone with you. Place it somewhere where you will have it in your energy field. Sodalite will also strengthen the immune system. When working with this stone, you should be sure to clean it often.

Topaz

Famous for its color, the topaz is clearer than glass and is highly prized (and expensive).

Blue topaz is called the "truth stone." This is good for blocks in the communications center (fifth chakra). It brings a new level of communication and releases old sadness. Topaz is a stone of exceptional optical qualities. The incredible clarity of topaz may be part of what gives it the qualities of truth.

White topaz helps you manifest abundance on a physical level. Program it and see. White topaz is less expensive than other topazes.

Yellow topaz is a true honey gold. It is stimulating to the third chakra and can break up creative blocks. This is a terrific tool for artists of all sorts.

Tourmalinated Quartz

Often tourmaline inclusions in quartz are of dark green or black tourmaline. This adds to the energy of the quartz by adding the attributes of the tourmaline to the quartz; the quartz also acts as an amplifier and disperses the energy of the tourmaline over a greater area.

Tourmaline

The tourmalines are a group of related minerals (each color having a distinct name of its own). They are all super conductors, and basically transmuters of energy. *Change* is the key word when describing the attributes of tourmaline. Tourmaline in its various colors brings energy through the heart. Please note that rubilite and indicolite are also tourmalines.

Black tourmaline (shorl) is currently being used as a stone of protection. It deflects negativity away from the wearer. This is an excellent stone to use in urban angst areas and in any situation where you feel pressure and negativity coming from sources outside yourself. This stone protects the emotions and can be carried, worn, or rubbed between the fingers when the need for protection is the greatest. Actually, this stone can be used *anywhere* on the body where protection is needed. It is a wonderful stone to wear over the heart under certain circumstances. I like to recommend using aquamarine, turquoise, or rose quartz with the black tourmaline; in many people the black tourmaline can emphasize their serious nature, and wearing a stone of joy can balance this out. This black stone does not absorb negative energy; it deflects it, so it does not need to be cleansed.

Green tourmaline also has protective aspects, especially in the darker shades. This does not act in the same deflective way as shorl, but it can relieve anxieties that arise internally. This stone is best activated when rubbed between the fingers. The green is a strong healing color and strengthens the whole system. It can be a real energy booster, both for the short term and the long term. This stone also attracts prosperity on a material plane. Green tourmaline helps to effect change in one's life as well as changing anxiety (and other negatives) into positive energy. This stone will balance the hormone system over time.

Watermelon tourmaline is typically green and pink tourmaline, although it can also be found as blue and pink, and blue or green and black. A watermelon crystal looks green, until it is cut across its grain. Then it reveals its pink inside, and you know why it's called watermelon. This pink-green mix can also form along the length of the crystal. One end of the crystal can be green, the other end pink. This is generally called a bicolored stone. Watermelon tourmaline is a heart center stone and a perfect balancer for the heart because of the green/red (pink) combination. These colors are considered the colors of Christ and indeed this stone is a stone that works on the frequency of unconditional love. It helps one release negative feelings of hatred, jealousy, and judgment, balancing the sexual chakra with the heart chakra.

Turquoise

The green to blue turquoise has long been used by the Tibetans and the American Indians. It can be used for ancestor connection by those persons drawn to it. This is a fifth-chakra stone and can open the lines of communication of what is stored in the heart. This stone inspires happiness and joy. It brings a deeper appreciation for nature. Turquoise often is found with large amounts of silver running through the stone. This would increase its emotional aspects. Long used as a healing stone, it continues to bring balance and general health.

GEMSTONE CHART

The following chart is a guide to help you find the colored stone that may help you in eradicating a particular disease or dis-ease. Clear quartz, of course, is always an appropriate stone to use, and can be used in place of any of the other stones because quartz can transmit any healing vibration you would like to transmit. The colored stones, however, seem to vibrate at more specific levels than the quartz. You might think of quartz crystal as a general practitioner and other stones as specialists.

The stones in the chart are just suggestions. If you feel that a particular stone not mentioned may be of benefit in treatment, by all means trust your own judgment. These suggestions are based on generalizations, and each person is a specific case with unique needs. These suggestions are based, also, on generalized causes of the physical manifestation.

If you think a particular stone will help you, it will. This is not wishful thinking or even a placebo effect. This is your intuition speaking to you. Trust yourself. Validate your feelings.

Gemstone Remedy Chart

In addition to seeing your doctor or other preferred therapist, you might want to try:

accidents . . . . carnelian, malachite, amethyst

aches and pains . . . . rose quartz, citrine, clear quartz

addictions to cigarettes, alcohol,
and drugs of all kinds . . . . sodalite, green tourmaline, amethyst

allergies . . . . blue lace agate, citrine, chrysocholla

anemia . . . . garnet, aquamarine, bloodstone

anorexia . . . . rose quartz, citrine, aquamarine

anxiety . . . . dark green tourmaline

arthritis . . . . azurite/malachite, gold

asthma (*see also* allergies) . . . . amethyst, aquamarine, green tourmaline

back discomforts:
- lower . . . . sodalite, malachite, green tourmaline
- middle . . . . bloodstone, amethyst, smoky quartz
- upper . . . . rose quartz, amethyst, bloodstone

continued on next page

baldness (from any cause)	bloodstone, hemimorphite
blood pressure (high and low)	obsidian, bloodstone, watermelon tourmaline
breasts (soreness, lumps, cysts)	green garnet, malachite, rose quartz, sodalite
bruises, burns	malachite, rose quartz
cancer	azurite/malachite, sodalite, rose quartz, malachite
colds, flu, viruses	sodalite, green tourmaline, citrine
constipation	black tourmaline, aquamarine, citrine
cramps:	
menstrual	smoky quartz, chrysocholla, clear quartz, aquamarine
stomach	citrine
diabetes	bloodstone, rose quartz, amethyst
eyes—nearsighted, farsighted, infections	green garnet, bloodstone
fatigue	clear quartz, amethyst, lepidolite
female discomforts (*see* breasts, cramps)	

headaches	green tourmaline, single- or double-terminated clear quartz directly on site of pain, ruby
heart ailments	amethyst, watermelon tourmaline, rose quartz or pink tourmaline, kunzite
hypoglycemia	clear quartz, aquamarine, bloodstone
indigestion	citrine, green tourmaline
insomnia	rose quartz, amethyst, smoky quartz
laryngitis	aquamarine, turquoise, blue topaz
nervousness	aquamarine, watermelon tourmaline, rose quartz, hematite or pyrite
sinus ailments	green garnet, amethyst, chrysocholla
skin—infections, rashes, acne	rutilated quartz, malachite, sodalite
spleen, liver, kidney	peridot, citrine
tonsillitis	rose quartz, yellow topaz, citrine

Ask and it will be given to you; seek and you will find; knock
and the door will be opened to you. For everyone who asks
receives; he who seeks finds; and to him who knocks, the
door will be opened.

Matthew 7:7–8